M208 Pure M

The Open University

AA3

Series

This publication forms part of an Open University course. Details of this and other Open University courses can be obtained from the Student Registration and Enquiry Service, The Open University, PO Box 197, Milton Keynes, MK7 6BJ, United Kingdom: tel. +44 (0)870 333 4340, e-mail general-enquiries@open.ac.uk

Alternatively, you may visit the Open University website at http://www.open.ac.uk where you can learn more about the wide range of courses and packs offered at all levels by The Open University.

To purchase a selection of Open University course materials, visit the webshop at www.ouw.co.uk, or contact Open University Worldwide, Michael Young Building, Walton Hall, Milton Keynes, MK7 6AA, United Kingdom, for a brochure: tel. +44 (0)1908 858785, fax +44 (0)1908 858787, e-mail ouwenq@open.ac.uk

The Open University, Walton Hall, Milton Keynes, MK7 6AA.

First published 2006.

Edited, designed and typeset by The Open University, using the Open University TEX System.

Printed and bound in the United Kingdom by Hobbs the Printers Limited, Brunel Road, Totton, Hampshire SO40 3WX.

ISBN 0 7492 0209 2

1.1

Contents

Introduction

The Ancient Greek philosopher and mathematician Zeno proposed a number of paradoxes of the infinite, which have intrigued succeeding generations. For example, Zeno claimed that it is impossible for an object to travel a given distance, since it must first travel half the distance, then half of the remaining distance, then half of what remains, and so on. There must always remain some distance left to travel, so the journey cannot be completed.

Zeno (c.490–420 BC) lived in Elea in southern Italy. His paradoxes included 'The Flying Arrow' and 'Achilles and the Tortoise'.

This paradox relies partly on the intuitive feeling that it is impossible to add up an infinite number of positive quantities and obtain a finite answer. However, the following illustration of the paradox suggests that such an infinite sum is plausible.

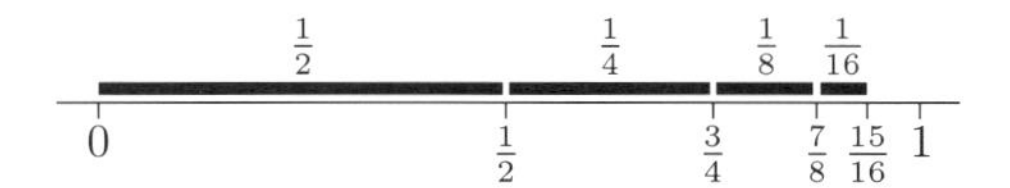

The distance from 0 to 1 can be split up into the infinite sequence of distances $\frac{1}{2}, \frac{1}{4}, \frac{1}{8}, \ldots$, so it seems reasonable to write

$$\frac{1}{2} + \frac{1}{4} + \frac{1}{8} + \frac{1}{16} + \cdots = 1.$$

This unit is devoted to the study of such expressions, which are called *infinite series*.

The following example shows that infinite series need to be treated with care. Suppose that it is possible to add up $2, 4, 8, \ldots$, and that the answer is s:

$$2 + 4 + 8 + \cdots = s.$$

If we multiply through by $\frac{1}{2}$, then we obtain

$$1 + 2 + 4 + 8 + \cdots = \tfrac{1}{2}s,$$

which can be written as $1 + s = \frac{1}{2}s$. Thus we deduce that $s = -2$, which is nonsense.

We can avoid reaching such absurd conclusions by performing arithmetic operations only with *convergent* infinite series; that is, those for which the corresponding sequence of successive *finite* sums is convergent.

In Section 1 we define this concept in terms of convergent sequences and give some examples. We also describe various properties which are common to all convergent series.

Section 2 is devoted to series with non-negative terms. We give several tests for the convergence of such series.

In Section 3 we deal with the much harder problem of determining whether a series is convergent when it has both positive and negative terms; the section ends with a general strategy for testing a given series for convergence.

In Section 4 we explain how e^x can be represented as an infinite series of powers of x. We use this representation to prove that the number e is irrational and also that $e^x e^y = e^{x+y}$.

Study guide

It is essential to study the sections in numerical order. Section 2 contains the audio section. Section 4 is for reading only; there are no exercises in this section.

This unit depends heavily on the ideas and results of Unit AA2; therefore, before studying this unit you should make sure that you understand Unit AA2, Sections 1, 2 and 3, and that you are familiar with Sections 4 and 5.

When working with series, we often need to determine whether a related sequence converges and, if so, the value of its limit. In such calculations, we do not always give as much detail as we did in Unit AA2. As usual, the amount of detail given in solutions to the examples and exercises in units indicates the amount you should give in solutions.

The video programme is a general one and can be watched at any time during your study of Sections 1–3 of the unit. It provides a useful overview of infinite series, as it touches on many of the key results in the unit. Among the topics covered are:

1. the convergence of the geometric series
$$\sum_{n=1}^{\infty} \frac{1}{2^n} = \frac{1}{2} + \frac{1}{4} + \frac{1}{8} + \frac{1}{16} + \cdots;$$
2. the convergence of the series
$$\sum_{n=1}^{\infty} \frac{1}{n^2} = 1 + \frac{1}{2^2} + \frac{1}{3^2} + \frac{1}{4^2} + \cdots;$$
3. the divergence of the series
$$\sum_{n=1}^{\infty} \frac{1}{n} = 1 + \frac{1}{2} + \frac{1}{3} + \frac{1}{4} + \cdots;$$
4. the fact that if a series converges, then its terms form a null sequence, but the converse statement does not hold;
5. the convergence of the series
$$\sum_{n=0}^{\infty} \frac{x^n}{n!} = 1 + x + \frac{x^2}{2!} + \frac{x^3}{3!} + \cdots, \quad \text{for } x \in \mathbb{R};$$
6. the equivalence of the definitions
$$e = \sum_{n=0}^{\infty} \frac{1}{n!} \quad \text{and} \quad e = \lim_{n\to\infty} (1 + 1/n)^n,$$
and, more generally, of the definitions
$$e^x = \sum_{n=0}^{\infty} \frac{x^n}{n!} \quad \text{and} \quad e^x = \lim_{n\to\infty} (1 + x/n)^n, \quad \text{for } x \in \mathbb{R}.$$

1 Introducing series

After working through this section, you should be able to:

(a) explain what is meant by a *convergent* series $\sum_{n=1}^{\infty} a_n$;
(b) write down the sum of a convergent *geometric* series;
(c) find the sums of certain *telescoping* series;
(d) use the Combination Rules for convergent series;
(e) use the Non-null Test to recognise certain *divergent* series.

1.1 What is a convergent series?

We begin by defining what is meant by the statement

$$\frac{1}{2}+\frac{1}{4}+\frac{1}{8}+\cdots=1.$$

Let s_n be the sum of the first n terms on the left-hand side. Then

$$s_1=\frac{1}{2},$$
$$s_2=\frac{1}{2}+\frac{1}{4}=\frac{3}{4},$$
$$s_3=\frac{1}{2}+\frac{1}{4}+\frac{1}{8}=\frac{7}{8}.$$

In general, using the formula for the sum of a finite geometric series with $a=r=\frac{1}{2}$, we obtain

The formula is
$$a+ar+\cdots+ar^{n-1}=\frac{a(1-r^n)}{1-r},$$
for $r\neq 1$.

$$s_n=\frac{1}{2}+\frac{1}{4}+\cdots+\left(\frac{1}{2}\right)^n=\frac{\frac{1}{2}(1-(\frac{1}{2})^n)}{1-\frac{1}{2}}=1-(\tfrac{1}{2})^n.$$

The sequence $\{(\frac{1}{2})^n\}$ is a null sequence, so

$$\lim_{n\to\infty} s_n = 1-\lim_{n\to\infty}(\tfrac{1}{2})^n=1.$$

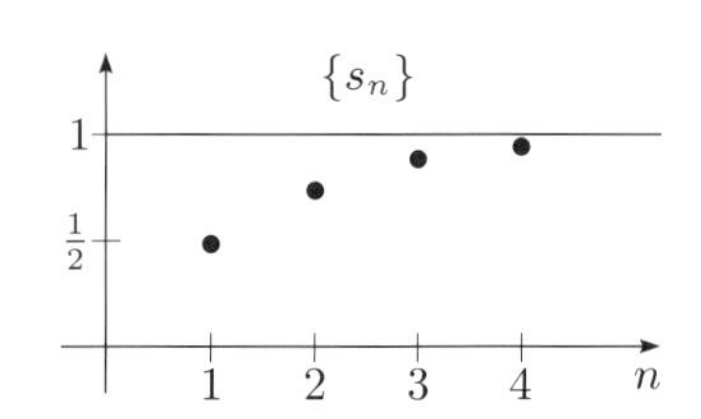

It is this precise mathematical statement that $s_n\to 1$ as $n\to\infty$ which justifies our informal statement in the Introduction that

$$\frac{1}{2}+\frac{1}{4}+\frac{1}{8}+\cdots=1.$$

We use this approach to define a *convergent infinite series*.

> **Definition** Let $\{a_n\}$ be a sequence. Then the expression
>
> $$a_1+a_2+a_3+\cdots$$
>
> is an **infinite series**, or simply a **series**. We call a_n the **nth term** of the series.
>
> The **nth partial sum** of this series is
>
> $$s_n=a_1+a_2+\cdots+a_n. \qquad (1.1)$$

The behaviour of the infinite series

$$a_1+a_2+a_3+\cdots$$

is determined by the behaviour of $\{s_n\}$, its sequence of partial sums.

> **Definition** The series
>
> $$a_1 + a_2 + a_3 + \cdots$$
>
> is **convergent** with **sum** s if its sequence $\{s_n\}$ of partial sums converges to s. In this case, the series **converges to** s and we write
>
> $$a_1 + a_2 + a_3 + \cdots = s.$$
>
> The series **diverges**, or is **divergent**, if the sequence $\{s_n\}$ diverges.

Thus we can prove results about a series by applying known results about sequences to its partial sums $\{s_n\}$.

Example 1.1 For each of the following series, calculate the nth partial sum, and determine whether the series is convergent or divergent.

(a) $1 + 1 + 1 + \cdots$ (b) $\frac{1}{3} + (\frac{1}{3})^2 + (\frac{1}{3})^3 + \cdots$ (c) $2 + 4 + 8 + \cdots$

Solution

(a) In this case,

$$s_n = 1 + 1 + \cdots + 1 = n.$$

The sequence $\{n\}$ tends to infinity, so this series is divergent.

(b) Using the formula for summing a finite geometric series with $a = r = \frac{1}{3}$, we obtain

$$s_n = \tfrac{1}{3} + (\tfrac{1}{3})^2 + (\tfrac{1}{3})^3 + \cdots + (\tfrac{1}{3})^n = \tfrac{1}{3}\frac{(1 - (\frac{1}{3})^n)}{1 - \frac{1}{3}} = \tfrac{1}{2}(1 - (\tfrac{1}{3})^n).$$

Since $\{(\frac{1}{3})^n\}$ is a basic null sequence,

$$\lim_{n\to\infty} s_n = \tfrac{1}{2},$$

so this series is convergent, with sum $\frac{1}{2}$.

Recall that the basic null sequences are:

$\left\{\frac{1}{n^p}\right\}$, for $p > 0$;

$\{c^n\}$, for $|c| < 1$;

$\{n^p c^n\}$, for $p > 0, |c| < 1$;

$\left\{\frac{c^n}{n!}\right\}$, for $c \in \mathbb{R}$;

$\left\{\frac{n^p}{n!}\right\}$, for $p > 0$.

See Unit AA2, Section 2.

(c) In this case the formula gives

$$s_n = 2 + 4 + 8 + \cdots + 2^n = 2\left(\frac{1 - 2^n}{1 - 2}\right) = 2^{n+1} - 2.$$

Now $2^{n+1} - 2 \to \infty$ as $n \to \infty$, so this series is divergent. ■

Remark It is important to distinguish between a series $a_1 + a_2 + \cdots$ and its sequence of terms $\{a_n\}$. The series $a_1 + a_2 + \cdots$ is an alternative notation for the sequence $\{s_n\}$ of partial sums, given by equation (1.1). It may be helpful to think of the series $a_1 + a_2 + \cdots$ as a 'snake' which

- has length s_n on its nth birthday, and
- grows by the length a_n in its nth year.

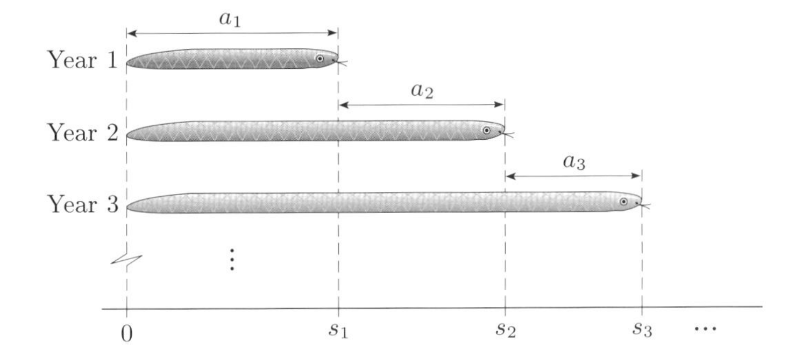

This picture has its limitations, however; for example, it assumes that the snake lives forever. Also the terms a_n need not be positive, so the snake may shrink or even have negative length!

Note that changing, deleting or adding a *finite* number of terms does not affect the convergence of a series, but may affect its sum. For example, the series

$$1+2+3+4+5+\frac{1}{2}+\frac{1}{4}+\frac{1}{8}+\cdots$$

is convergent with sum $1+2+3+4+5+1=16$.

Sigma notation

Next we explain how to use *sigma notation* to represent infinite series.

A finite sum such as

$$a_1+a_2+\cdots+a_{10}$$

can be represented using sigma notation as $\sum_{n=1}^{10} a_n$. This notation can be adapted to represent infinite series

$$\sum_{n=1}^{\infty} a_n = a_1+a_2+a_3+\cdots,$$

read as 'sigma, $n=1$ to infinity, a_n', or 'the sum from $n=1$ to infinity of a_n'.

For example,

$$\sum_{n=1}^{\infty}\frac{1}{n}=1+\frac{1}{2}+\frac{1}{3}+\cdots.$$

Remarks

1. When using sigma notation to represent the nth partial sum s_n of a series, we use another letter for the subscript of the terms to avoid n having two different meanings in the same expression; for example,

$$s_n=a_1+a_2+\cdots+a_n=\sum_{k=1}^{n} a_k.$$

The letters i,j,k,l,m,n,p and q are used for subscript variables.

For the above series,

$$s_n=1+\frac{1}{2}+\cdots+\frac{1}{n}=\sum_{k=1}^{n}\frac{1}{k}.$$

2. If a series begins with a term other than a_1, then we adapt this notation appropriately; for example,

$$\sum_{n=0}^{\infty} a_n = a_0+a_1+a_2+\cdots \quad \text{or} \quad \sum_{n=3}^{\infty} a_n = a_3+a_4+a_5+\cdots.$$

For such series, the nth partial sum s_n is obtained by adding all the terms up to and including a_n. For example, the nth partial sum of

$$\sum_{n=0}^{\infty} a_n \quad \text{is} \quad s_n=a_0+a_1+\cdots+a_n=\sum_{k=0}^{n} a_k.$$

For example, we write $\sum_{n=0}^{\infty} r^n$ for the series $1+r+r^2+\cdots$.

3. An alternative layout for sigma notation is $\sum_{n=1}^{\infty} a_n$.

4. We sometimes use the simpler notation $\sum a_n$ to denote a general infinite series with terms a_n.

Exercise 1.1 For each of the following series, calculate the nth partial sum s_n and determine whether the series is convergent or divergent.

(a) $\sum_{n=1}^{\infty}\left(-\frac{1}{3}\right)^n$ (b) $\sum_{n=1}^{\infty}(-1)^n$ (c) $\sum_{n=0}^{\infty}\left(\frac{1}{2}\right)^n$

The series considered so far in this section are all geometric series. The **(infinite) geometric series** with first term a and common ratio r is

$$\sum_{n=0}^{\infty} ar^n = a + ar + ar^2 + \cdots.$$

The following theorem enables us to decide whether any given geometric series is convergent or divergent.

Geometric series

(a) If $|r| < 1$, then $\sum_{n=0}^{\infty} ar^n$ is convergent, with sum $\dfrac{a}{1-r}$.

(b) If $|r| \geq 1$ and $a \neq 0$, then $\sum_{n=0}^{\infty} ar^n$ is divergent.

Proof

(a) If $r \neq 1$, then the nth partial sum s_n is given by

$$s_n = a + ar + ar^2 + \cdots + ar^n = \frac{a(1-r^{n+1})}{1-r}. \qquad (1.2)$$

Now, if $|r| < 1$, then $\{r^n\}$ is a basic null sequence, so

$$\begin{aligned}\lim_{n\to\infty} s_n &= \lim_{n\to\infty} \frac{a(1-r^{n+1})}{1-r} \\ &= \frac{a}{1-r}\left(1 - \lim_{n\to\infty} r^{n+1}\right) = \frac{a}{1-r},\end{aligned}$$

by the Combination Rules for sequences. Thus, if $|r| < 1$, then

See Unit AA2, Section 3.

$$\sum_{n=0}^{\infty} ar^n \text{ is convergent, with sum } a/(1-r).$$

(b) We know already that a geometric series with $r = \pm 1$ is divergent.

See Example 1.1(a) and Exercise 1.1(b).

If $|r| > 1$, then equation (1.2) holds and $\{|r|^{n+1}\}$ tends to infinity, so the sequence $\{s_n\}$ is unbounded and hence divergent. Thus, if $|r| \geq 1$, then $\sum_{n=0}^{\infty} ar^n$ is divergent. ■

1.2 Telescoping series

Geometric series are easy to deal with because there is a formula for the nth partial sum s_n. The next exercise deals with another series for which we can calculate a formula for s_n.

Exercise 1.2 Calculate the first four partial sums of the following series, giving your answers as fractions:

$$\sum_{n=1}^{\infty} \frac{1}{n(n+1)} = \frac{1}{1\times 2} + \frac{1}{2\times 3} + \frac{1}{3\times 4} + \cdots.$$

The partial sums obtained in Exercise 1.2 suggest the general formula

$$s_n = \frac{1}{1\times 2} + \frac{1}{2\times 3} + \cdots + \frac{1}{n(n+1)} = \frac{n}{n+1}.$$

This formula can be proved by using the identity

$$\frac{1}{n(n+1)} = \frac{1}{n} - \frac{1}{n+1}, \quad \text{for } n = 1, 2, \ldots,$$

which implies that

$$\begin{aligned} s_n &= \frac{1}{1\times 2} + \frac{1}{2\times 3} + \cdots + \frac{1}{(n-1)n} + \frac{1}{n(n+1)} \\ &= \left(\frac{1}{1} - \frac{1}{2}\right) + \left(\frac{1}{2} - \frac{1}{3}\right) + \cdots + \left(\frac{1}{n-1} - \frac{1}{n}\right) + \left(\frac{1}{n} - \frac{1}{n+1}\right) \\ &= 1 - \frac{1}{n+1} = \frac{n}{n+1}. \end{aligned}$$

Cancellation of the adjacent terms

$-\frac{1}{2}, \frac{1}{2}, -\frac{1}{3}, \frac{1}{3}, \ldots,$

explains why this series is said to be *telescoping*.

Since

$$\lim_{n\to\infty} s_n = \lim_{n\to\infty} \frac{n}{n+1} = 1,$$

we deduce that the given series is convergent, with sum

$$\sum_{n=1}^{\infty} \frac{1}{n(n+1)} = 1.$$

Exercise 1.3 Find the nth partial sum s_n of $\sum_{n=1}^{\infty} \frac{1}{n(n+2)}$, using the identity

$$\frac{2}{n(n+2)} = \frac{1}{n} - \frac{1}{n+2}, \quad \text{for } n = 1, 2, \ldots.$$

Deduce that this series is convergent and find its sum.

1.3 Combination Rules for convergent series

In the Introduction we saw that performing arithmetic operations on the divergent series $2 + 4 + 8 + \cdots$ can lead to absurd conclusions. However, we can perform arithmetic operations on convergent series. The following result shows that there are Combination Rules for convergent series, which follow directly from the Combination Rules for sequences.

Combination Rules Suppose that $\sum_{n=1}^{\infty} a_n = s$ and $\sum_{n=1}^{\infty} b_n = t$. Then

Sum Rule $\sum_{n=1}^{\infty} (a_n + b_n) = s + t$;

Multiple Rule $\sum_{n=1}^{\infty} \lambda a_n = \lambda s$, for $\lambda \in \mathbb{R}$.

Proof Consider the sequences of partial sums $\{s_n\}$ and $\{t_n\}$, where

$$s_n = \sum_{k=1}^{n} a_k \quad \text{and} \quad t_n = \sum_{k=1}^{n} b_k.$$

We know that $s_n \to s$ as $n \to \infty$ and $t_n \to t$ as $n \to \infty$.

Sum Rule The nth partial sum of the series $\sum_{n=1}^{\infty}(a_n + b_n)$ is

$$\begin{aligned}\sum_{k=1}^{n}(a_k + b_k) &= (a_1 + b_1) + (a_2 + b_2) + \cdots + (a_n + b_n)\\ &= (a_1 + a_2 + \cdots + a_n) + (b_1 + b_2 + \cdots + b_n)\\ &= s_n + t_n.\end{aligned}$$

By the Sum Rule for sequences,

$$\lim_{n\to\infty}(s_n + t_n) = \lim_{n\to\infty} s_n + \lim_{n\to\infty} t_n = s + t,$$

so the sequence $\{s_n + t_n\}$ of partial sums of $\sum_{n=1}^{\infty}(a_n + b_n)$ has limit $s + t$.

Hence this series is convergent and

$$\sum_{n=1}^{\infty}(a_n + b_n) = s + t.$$

Multiple Rule The nth partial sum of the series $\sum_{n=1}^{\infty} \lambda a_n$ is

$$\begin{aligned}\sum_{k=1}^{n} \lambda a_k &= \lambda a_1 + \lambda a_2 + \cdots + \lambda a_n\\ &= \lambda(a_1 + a_2 + \cdots + a_n)\\ &= \lambda s_n.\end{aligned}$$

By the Multiple Rule for sequences,

$$\lim_{n\to\infty}(\lambda s_n) = \lambda \lim_{n\to\infty} s_n = \lambda s,$$

so the sequence $\{\lambda s_n\}$ of partial sums of $\sum_{n=1}^{\infty} \lambda a_n$ has limit λs. Hence this series is convergent and

$$\sum_{n=1}^{\infty} \lambda a_n = \lambda s. \quad \blacksquare$$

Example 1.2 Prove that the following series is convergent and calculate its sum:

$$\sum_{n=1}^{\infty}\left(\frac{1}{2^n} + \frac{3}{n(n+1)}\right).$$

Solution We know that

$$\sum_{n=1}^{\infty} \frac{1}{2^n} \text{ is a convergent geometric series, with sum 1,}$$

and that

$$\sum_{n=1}^{\infty} \frac{1}{n(n+1)} \text{ is convergent, with sum 1.}$$

See Subsection 1.2.

Hence, by the Sum and the Multiple Rules,

$$\sum_{n=1}^{\infty} \left(\frac{1}{2^n} + \frac{3}{n(n+1)}\right) \text{ is convergent, with sum } 1 + (3 \times 1) = 4. \quad \blacksquare$$

Exercise 1.4 Prove that the following series is convergent and calculate its sum:

$$\sum_{n=1}^{\infty} \left(\left(-\tfrac{3}{4}\right)^n - \frac{2}{n(n+1)}\right).$$

1.4 Non-null Test

For all the infinite series we have so far considered, it is possible to derive a simple formula for the nth partial sum. For many series, however, this is difficult or even impossible.

Nevertheless, it may still be possible to decide whether such series are convergent or divergent by applying various tests. Our first test arises from the following result.

> **Theorem 1.1** If $\sum_{n=1}^{\infty} a_n$ is a convergent series, then its sequence of terms $\{a_n\}$ is a null sequence.

Proof Let $s_n = \sum_{k=1}^{n} a_k$ denote the nth partial sum of $\sum_{n=1}^{\infty} a_n$. Because $\sum_{n=1}^{\infty} a_n$ is convergent, we know that $\{s_n\}$ is convergent, with limit s say.

We want to deduce that $\{a_n\}$ is null. To do this, we note that

$$a_n = s_n - s_{n-1}, \quad \text{for } n \geq 2,$$

by equation (1.1). Thus, by the Combination Rules,

$$\begin{aligned}\lim_{n\to\infty} a_n = \lim_{n\to\infty} (s_n - s_{n-1}) &= \lim_{n\to\infty} s_n - \lim_{n\to\infty} s_{n-1} \\ &= s - s = 0.\end{aligned}$$

The sequence $\{s_{n-1}\}_2^{\infty}$ is the same as the sequence $\{s_n\}_1^{\infty}$, so

$$s_{n-1} \to s \text{ as } n \to \infty.$$

Hence $\{a_n\}$ is a null sequence, as required. $\blacksquare$

The following test for divergence is an immediate corollary of Theorem 1.1.

Corollary Non-null Test

If $\{a_n\}$ is not a null sequence, then $\sum_{n=1}^{\infty} a_n$ is divergent.

The main virtue of the Non-null Test is its ease of application. For example, it enables us to decide immediately that the two series

$$\sum_{n=1}^{\infty}(-1)^n \quad \text{and} \quad \sum_{n=1}^{\infty} 2^n/n \quad \text{are divergent}$$

because the corresponding sequences of terms

$$\{(-1)^n\} \quad \text{and} \quad \{2^n/n\} \quad \text{are not null.}$$

We have
$|(-1)^n| \to 1$ as $n \to \infty$,
and
$|2^n/n| \to \infty$ as $n \to \infty$.
The latter holds by the Reciprocal Rule; see Unit AA2, Section 4.

Finally, we discuss how to show that a given sequence $\{a_n\}$ *does not* tend to zero. We know that if $\{a_n\}$ is a null sequence, then $\{|a_n|\}$ is also null, as are all subsequences of $\{|a_n|\}$. This leads to the following strategy.

See Unit AA2, Theorem 4.2(a).

Strategy 1.1

To show that $\sum_{n=1}^{\infty} a_n$ is divergent using the Non-null Test

EITHER

1. show that $\{|a_n|\}$ has a convergent subsequence with non-zero limit,

OR

2. show that $\{|a_n|\}$ has a subsequence which tends to infinity.

Often, when we apply the Non-null Test, the sequence $\{|a_n|\}$ itself tends to a non-zero limit or to infinity, as in the two examples after the Non-null Test.

Exercise 1.5 Prove that $\displaystyle\sum_{n=1}^{\infty} \frac{(-1)^{n+1}n^2}{2n^2+1}$ is divergent.

Note that the converse of the Non-null Test is *false*. If the sequence $\{a_n\}$ is null, then it is not necessarily true that the series $\sum_{n=1}^{\infty} a_n$ is convergent. For example, the sequence $\{1/n\}$ is null, but the series

$$\sum_{n=1}^{\infty} \frac{1}{n} = 1 + \frac{1}{2} + \frac{1}{3} + \cdots \quad \text{is divergent.}$$

We prove this rather surprising fact in Subsection 2.1.

So you can *never* use the Non-null Test to prove that a series is convergent.

Further exercises

Exercise 1.6 Prove that $\sum_{n=0}^{\infty} \left(-\frac{4}{5}\right)^n$ is convergent and find its sum.

Exercise 1.7 Prove that

$$\frac{1}{2n-1} - \frac{1}{2n+1} = \frac{2}{4n^2-1}, \quad \text{for } n = 1, 2, \ldots,$$

and deduce that

$$\sum_{n=1}^{\infty} \frac{1}{4n^2-1} = \tfrac{1}{2}.$$

Exercise 1.8 Determine the behaviour of the following series.

(a) $\displaystyle\sum_{n=1}^{\infty} \left(\left(\tfrac{4}{5}\right)^n + \frac{4}{n(n+2)} \right)$ (b) $\displaystyle\sum_{n=1}^{\infty} \left(1 + \left(\tfrac{1}{2}\right)^n\right)$

2 Series with non-negative terms

After working through this section, you should be able to:

(a) explain why $\displaystyle\sum_{n=1}^{\infty} 1/n$ is divergent and $\displaystyle\sum_{n=1}^{\infty} 1/n^2$ is convergent;
(b) use the Comparison Test and the Limit Comparison Test;
(c) use the Ratio Test;
(d) recognise and use *basic* series.

In this section we consider only series $\displaystyle\sum_{n=1}^{\infty} a_n$ with non-negative terms. In other words, we assume that

$$a_n \geq 0, \quad \text{for } n = 1, 2, \ldots.$$

It follows that the partial sums of $\displaystyle\sum_{n=1}^{\infty} a_n$, given by

$$\begin{aligned} s_1 &= a_1 \\ s_2 &= a_1 + a_2 \\ &\vdots \\ s_n &= a_1 + a_2 + \cdots + a_n, \end{aligned}$$

form an *increasing* sequence (as for the snake shown on page 7).

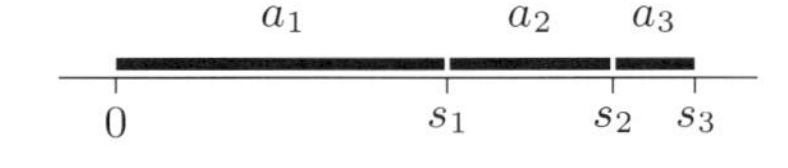

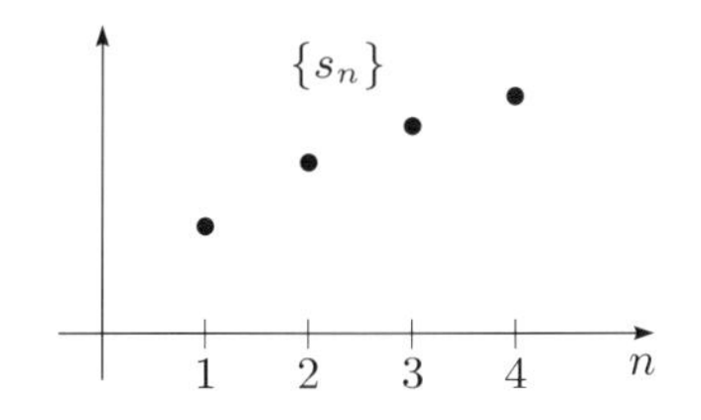

The fact that the sequence $\{s_n\}$ of partial sums is increasing makes it easier to deal with series having non-negative terms. If we can prove that $\{s_n\}$ is bounded above, then $\{s_n\}$ is convergent, by the Monotone Convergence Theorem, so $\displaystyle\sum_{n=1}^{\infty} a_n$ is convergent.

See Unit AA2, Section 5.

2.1 Tests for convergence

Before starting the audio you should try the following exercise.

Exercise 2.1 Use your calculator to find the first eight partial sums of each of the following series (giving your answers to 2 decimal places) and plot your answers on a sequence diagram:

$$\sum_{n=1}^{\infty} \frac{1}{n} = 1 + \frac{1}{2} + \frac{1}{3} + \cdots; \qquad \sum_{n=1}^{\infty} \frac{1}{n^2} = 1 + \frac{1}{2^2} + \frac{1}{3^2} + \cdots.$$

In the audio frames we give several tests for the convergence of series with non-negative terms. In some of these tests we find the limits of sequences whose terms are quotients by using the Combination Rules for sequences. When these quotients involve factorials, they can often be simplified by cancellations such as

$$\begin{aligned}\frac{(2n+2)!}{(2n)!} &= \frac{(2n+2)(2n+1)(2n)(2n-1)\cdot \cdots \cdot 2\cdot 1}{(2n)(2n-1)\cdot \cdots \cdot 2\cdot 1}\\ &= (2n+2)(2n+1).\end{aligned}$$

Listen to the audio as you work through the frames.

Audio

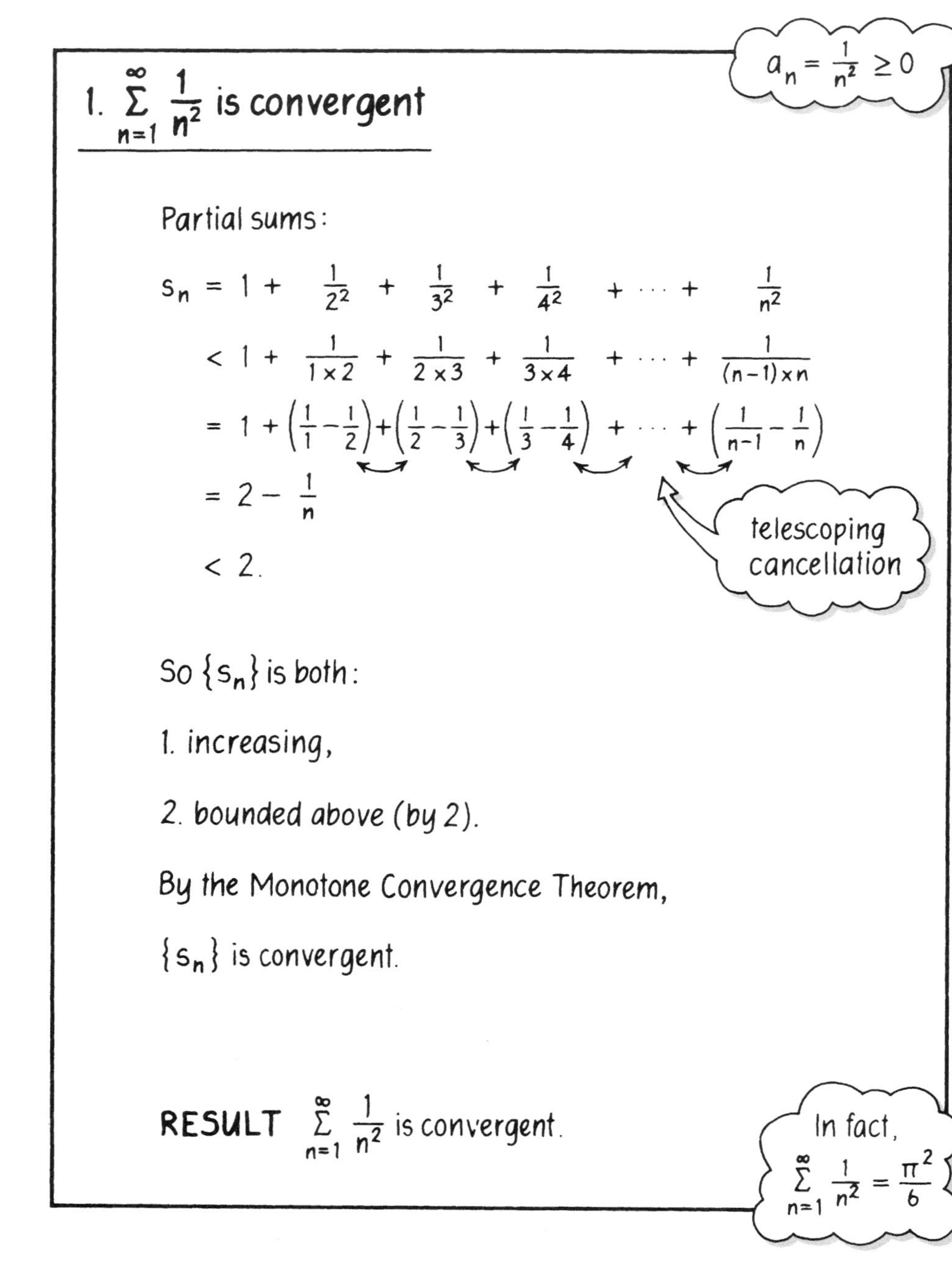

1. $\sum_{n=1}^{\infty} \frac{1}{n^2}$ is convergent

Partial sums:

$$
\begin{aligned}
s_n &= 1 + \frac{1}{2^2} + \frac{1}{3^2} + \frac{1}{4^2} + \cdots + \frac{1}{n^2} \\
&< 1 + \frac{1}{1 \times 2} + \frac{1}{2 \times 3} + \frac{1}{3 \times 4} + \cdots + \frac{1}{(n-1) \times n} \\
&= 1 + \left(\frac{1}{1} - \frac{1}{2}\right) + \left(\frac{1}{2} - \frac{1}{3}\right) + \left(\frac{1}{3} - \frac{1}{4}\right) + \cdots + \left(\frac{1}{n-1} - \frac{1}{n}\right) \\
&= 2 - \frac{1}{n} \\
&< 2.
\end{aligned}
$$

So $\{s_n\}$ is both:

1. increasing,

2. bounded above (by 2).

By the Monotone Convergence Theorem,

$\{s_n\}$ is convergent.

RESULT $\sum_{n=1}^{\infty} \frac{1}{n^2}$ is convergent.

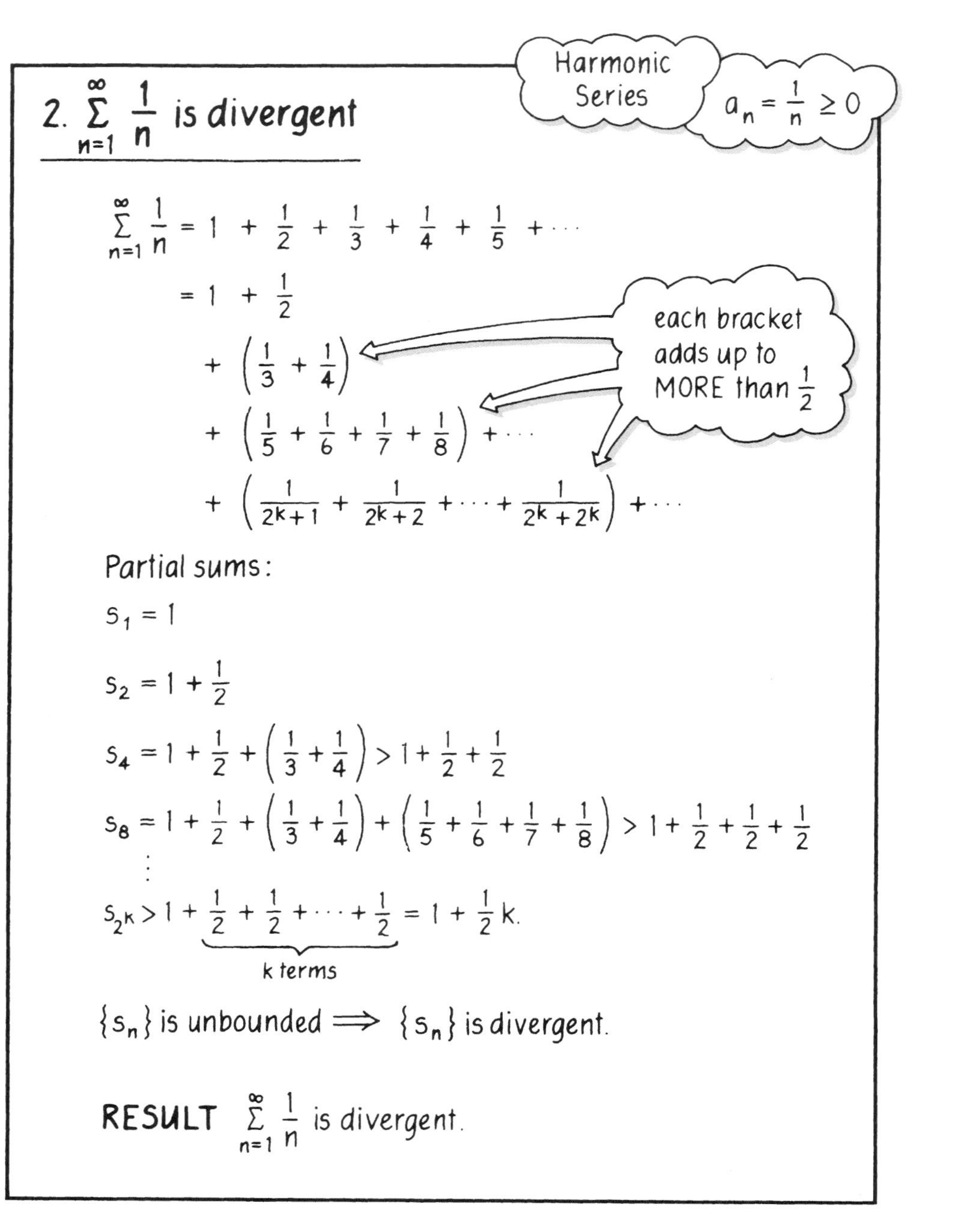

2. $\sum_{n=1}^{\infty} \frac{1}{n}$ is divergent

$$
\begin{aligned}
\sum_{n=1}^{\infty} \frac{1}{n} &= 1 + \frac{1}{2} + \frac{1}{3} + \frac{1}{4} + \frac{1}{5} + \cdots \\
&= 1 + \frac{1}{2} \\
&\quad + \left(\frac{1}{3} + \frac{1}{4}\right) \\
&\quad + \left(\frac{1}{5} + \frac{1}{6} + \frac{1}{7} + \frac{1}{8}\right) + \cdots \\
&\quad + \left(\frac{1}{2^k+1} + \frac{1}{2^k+2} + \cdots + \frac{1}{2^k+2^k}\right) + \cdots
\end{aligned}
$$

Partial sums:

$s_1 = 1$

$s_2 = 1 + \frac{1}{2}$

$s_4 = 1 + \frac{1}{2} + \left(\frac{1}{3} + \frac{1}{4}\right) > 1 + \frac{1}{2} + \frac{1}{2}$

$s_8 = 1 + \frac{1}{2} + \left(\frac{1}{3} + \frac{1}{4}\right) + \left(\frac{1}{5} + \frac{1}{6} + \frac{1}{7} + \frac{1}{8}\right) > 1 + \frac{1}{2} + \frac{1}{2} + \frac{1}{2}$

$\vdots$

$s_{2^k} > 1 + \underbrace{\frac{1}{2} + \frac{1}{2} + \cdots + \frac{1}{2}}_{k \text{ terms}} = 1 + \frac{1}{2}k.$

$\{s_n\}$ is unbounded $\Longrightarrow$ $\{s_n\}$ is divergent.

RESULT $\sum_{n=1}^{\infty} \frac{1}{n}$ is divergent.

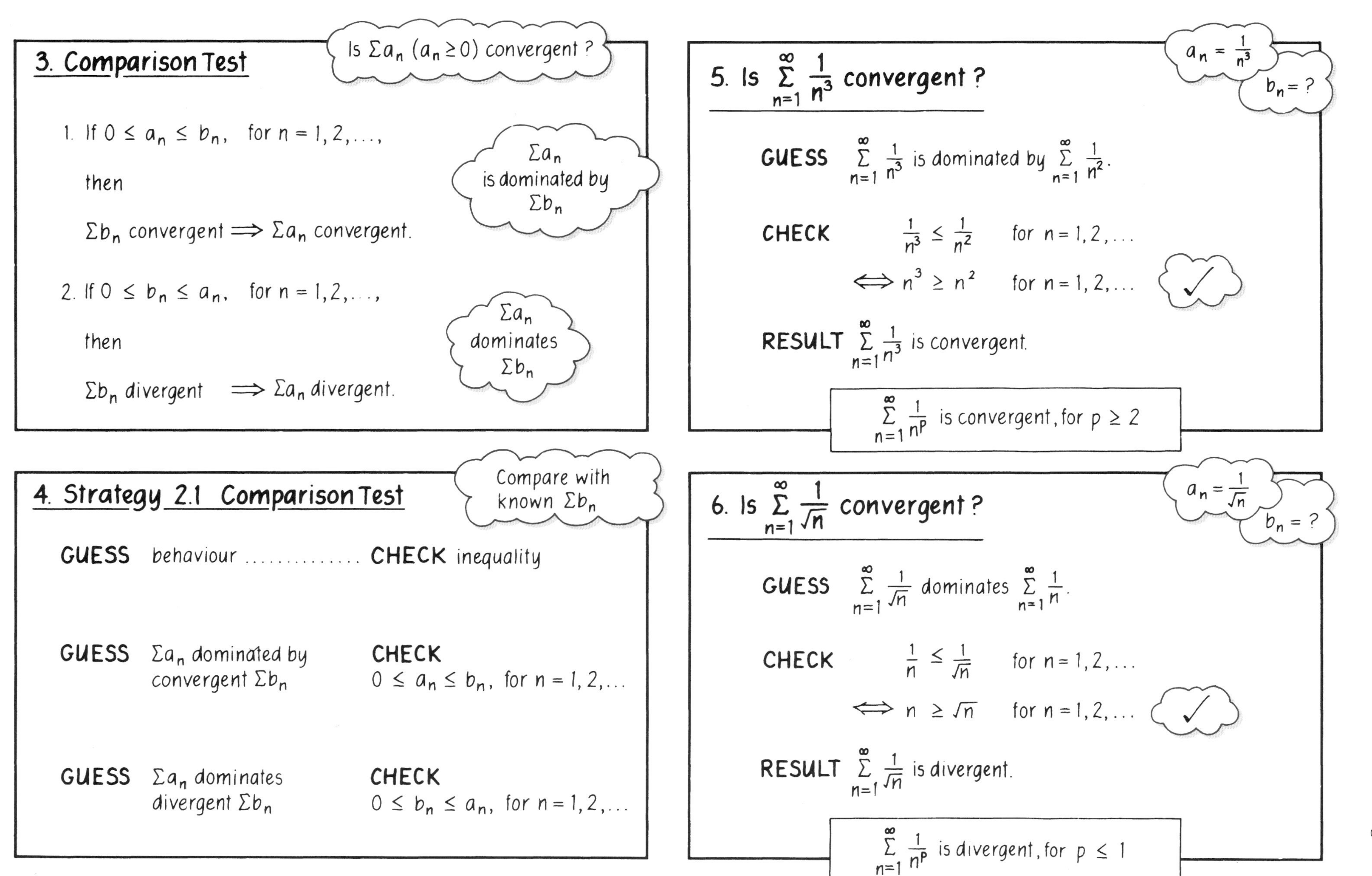
3. Comparison Test
Is Σa_n $(a_n \geq 0)$ convergent?
1. If $0 \leq a_n \leq b_n$, for $n = 1, 2, \ldots$,
then
Σb_n convergent $\Longrightarrow$ Σa_n convergent.
Σa_n is dominated by Σb_n
2. If $0 \leq b_n \leq a_n$, for $n = 1, 2, \ldots$,
then
Σb_n divergent $\Longrightarrow$ Σa_n divergent.
Σa_n dominates Σb_n
4. Strategy 2.1 Comparison Test
Compare with known Σb_n
GUESS behaviour .... CHECK inequality
GUESS Σa_n dominated by convergent Σb_n
CHECK $0 \leq a_n \leq b_n$, for $n = 1, 2, \ldots$
GUESS Σa_n dominates divergent Σb_n
CHECK $0 \leq b_n \leq a_n$, for $n = 1, 2, \ldots$
5. Is $\sum_{n=1}^{\infty} \frac{1}{n^3}$ convergent?
$a_n = \frac{1}{n^3}$
$b_n = ?$
GUESS $\sum_{n=1}^{\infty} \frac{1}{n^3}$ is dominated by $\sum_{n=1}^{\infty} \frac{1}{n^2}$.
CHECK $\frac{1}{n^3} \leq \frac{1}{n^2}$ for $n = 1, 2, \ldots$
$\Longleftrightarrow n^3 \geq n^2$ for $n = 1, 2, \ldots$
✓
RESULT $\sum_{n=1}^{\infty} \frac{1}{n^3}$ is convergent.
$\sum_{n=1}^{\infty} \frac{1}{n^p}$ is convergent, for $p \geq 2$
6. Is $\sum_{n=1}^{\infty} \frac{1}{\sqrt{n}}$ convergent?
$a_n = \frac{1}{\sqrt{n}}$
$b_n = ?$
GUESS $\sum_{n=1}^{\infty} \frac{1}{\sqrt{n}}$ dominates $\sum_{n=1}^{\infty} \frac{1}{n}$.
CHECK $\frac{1}{n} \leq \frac{1}{\sqrt{n}}$ for $n = 1, 2, \ldots$
$\Longleftrightarrow n \geq \sqrt{n}$ for $n = 1, 2, \ldots$
✓
RESULT $\sum_{n=1}^{\infty} \frac{1}{\sqrt{n}}$ is divergent.
$\sum_{n=1}^{\infty} \frac{1}{n^p}$ is divergent, for $p \leq 1$

7. Limit Comparison Test

Is $\sum a_n$ $(a_n > 0)$ convergent?

If $\sum a_n$ and $\sum b_n$ have positive terms and

$$\lim_{n\to\infty} \frac{a_n}{b_n} = L \neq 0,$$

then

1. $\sum b_n$ convergent $\Longrightarrow$ $\sum a_n$ convergent
2. $\sum b_n$ divergent $\Longrightarrow$ $\sum a_n$ divergent

8. Strategy 2.2

Compare with known $\sum b_n$

GUESS a_n behaves like b_n

CHECK $\frac{a_n}{b_n} \to L$ as $n \to \infty$, where $L \neq 0$.

9. Is $\sum_{n=1}^{\infty} \frac{1}{2\sqrt{n}+1}$ convergent?

$a_n = \frac{1}{2\sqrt{n}+1}$ $b_n = ?$

GUESS a_n behaves like $b_n = \frac{1}{\sqrt{n}}$.

CHECK $\frac{a_n}{b_n} = \frac{\sqrt{n}}{2\sqrt{n}+1} = \frac{1}{2+1/\sqrt{n}} \to \frac{1}{2}$ as $n \to \infty$

$\frac{1}{2} \neq 0$ ✓

RESULT $\sum_{n=1}^{\infty} \frac{1}{2\sqrt{n}+1}$ is divergent, since $\sum_{n=1}^{\infty} \frac{1}{\sqrt{n}}$ is.

10. Is $\sum_{n=1}^{\infty} \frac{n+5}{3n^4-n}$ convergent?

$a_n = \frac{n+5}{3n^4-n}$ $b_n = ?$

GUESS a_n behaves like $b_n = \frac{1}{n^3}$.

CHECK $\frac{a_n}{b_n} = \frac{(n+5)n^3}{3n^4-n} = \frac{n^4+5n^3}{3n^4-n} = \frac{1+5/n}{3-1/n^3}$

so

$\frac{a_n}{b_n} \to \frac{1}{3}$ as $n \to \infty$

$\frac{1}{3} \neq 0$ ✓

RESULT $\sum_{n=1}^{\infty} \frac{n+5}{3n^4-n}$ is convergent, since $\sum_{n=1}^{\infty} \frac{1}{n^3}$ is.

11. Exercise 2.2

Use the Comparison Test or the Limit Comparison Test to determine the behaviour of the following series:

(a) $\sum_{n=1}^{\infty} \frac{1}{n^3+n}$

(b) $\sum_{n=1}^{\infty} \frac{1}{n+\sqrt{n}}$

(c) $\sum_{n=1}^{\infty} \frac{n+4}{2n^3-n+1}$

(d) $\sum_{n=1}^{\infty} \frac{\cos^2(2n)}{n^3}$

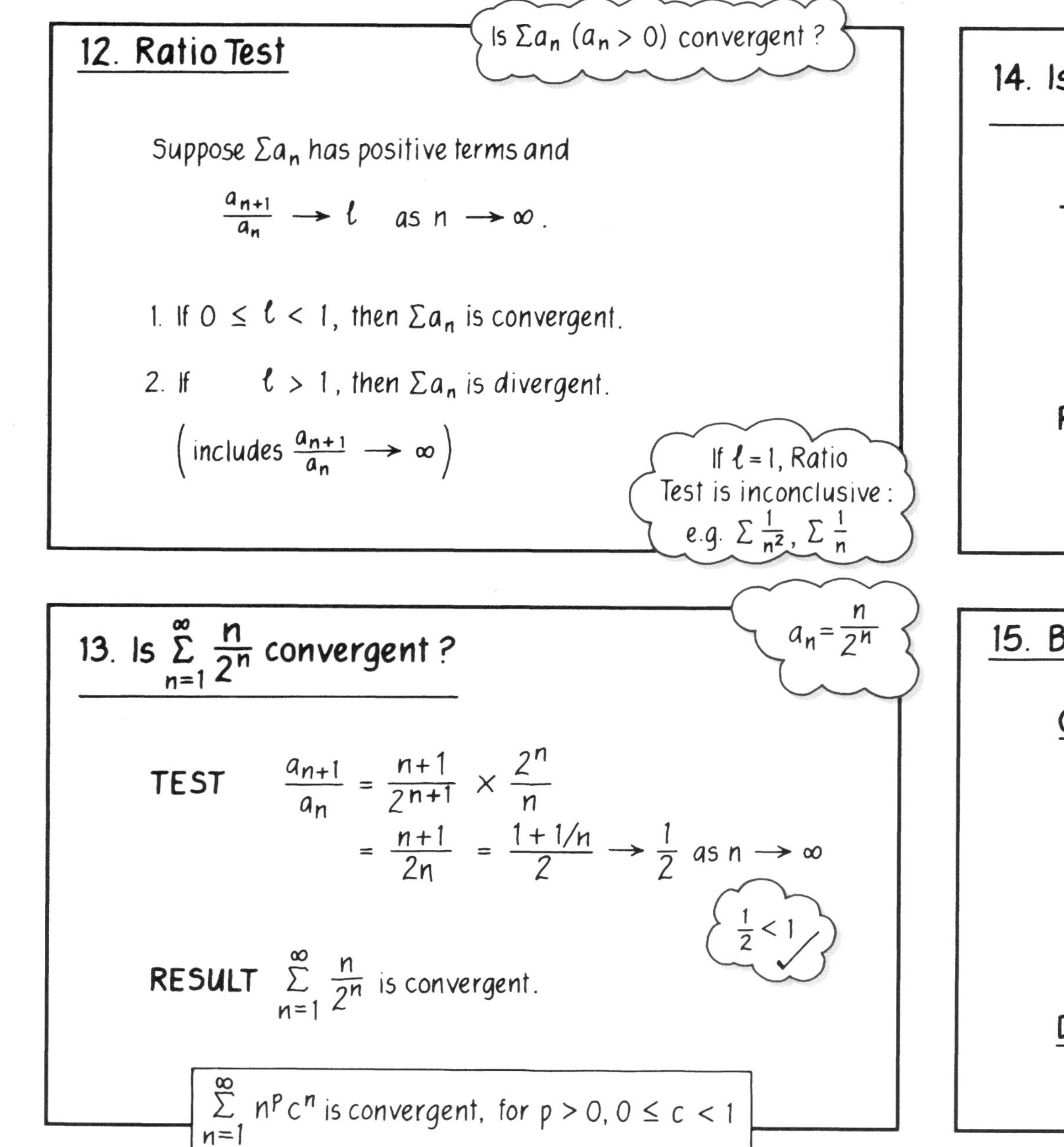

12. Ratio Test

Is Σa_n $(a_n > 0)$ convergent?

Suppose Σa_n has positive terms and

$$\frac{a_{n+1}}{a_n} \to \ell \quad \text{as } n \to \infty.$$

1. If $0 \le \ell < 1$, then Σa_n is convergent.
2. If $\ell > 1$, then Σa_n is divergent.

(includes $\frac{a_{n+1}}{a_n} \to \infty$)

If $\ell = 1$, Ratio Test is inconclusive: e.g. $\Sigma \frac{1}{n^2}$, $\Sigma \frac{1}{n}$

13. Is $\sum_{n=1}^{\infty} \frac{n}{2^n}$ convergent?

$a_n = \frac{n}{2^n}$

TEST $\quad \frac{a_{n+1}}{a_n} = \frac{n+1}{2^{n+1}} \times \frac{2^n}{n}$

$$= \frac{n+1}{2n} = \frac{1 + 1/n}{2} \to \frac{1}{2} \text{ as } n \to \infty$$

$\frac{1}{2} < 1$ ✓

RESULT $\quad \sum_{n=1}^{\infty} \frac{n}{2^n}$ is convergent.

> $\sum_{n=1}^{\infty} n^p c^n$ is convergent, for $p > 0, 0 \le c < 1$

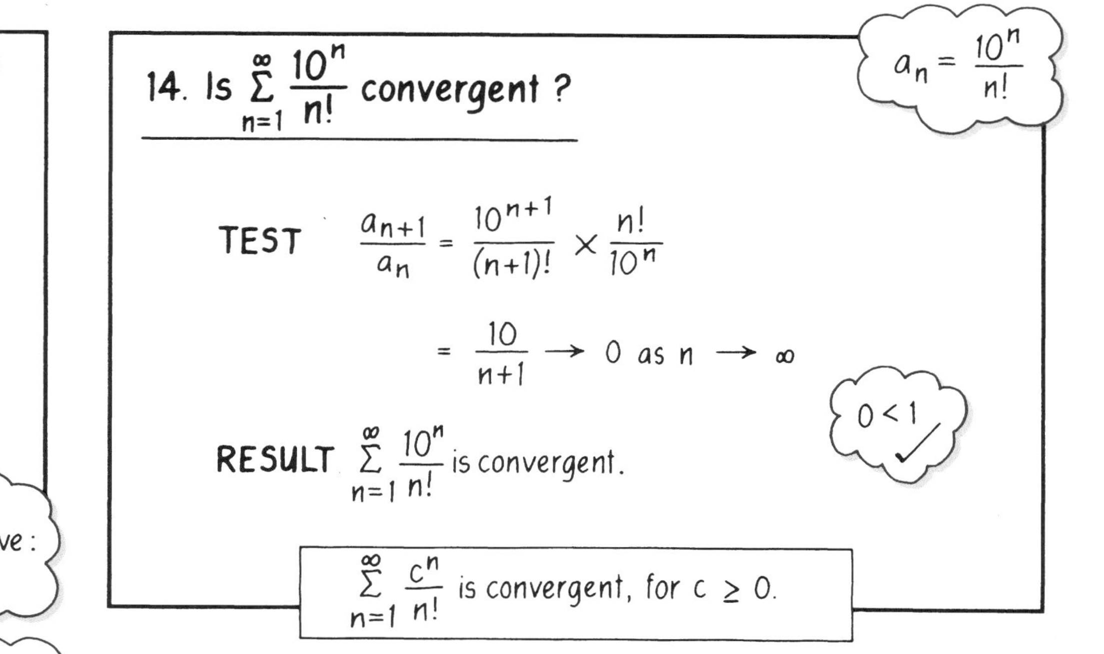

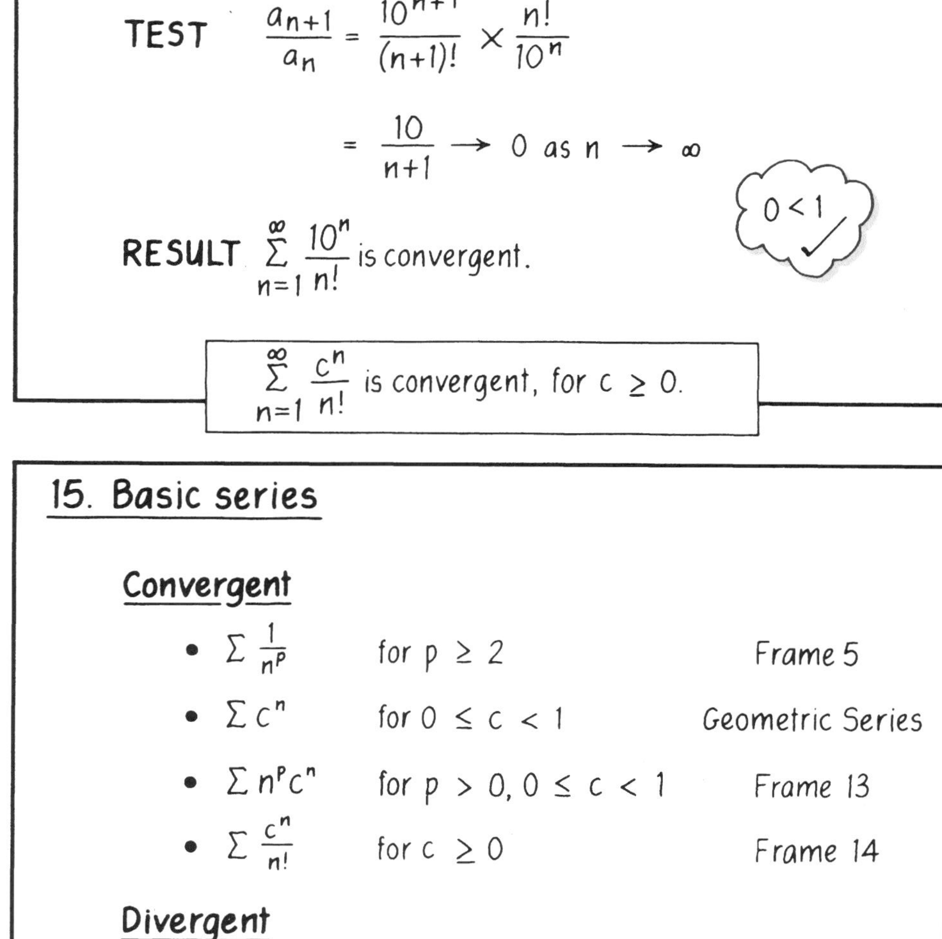

14. Is $\sum_{n=1}^{\infty} \frac{10^n}{n!}$ convergent?

$a_n = \frac{10^n}{n!}$

TEST $\quad \frac{a_{n+1}}{a_n} = \frac{10^{n+1}}{(n+1)!} \times \frac{n!}{10^n}$

$$= \frac{10}{n+1} \to 0 \text{ as } n \to \infty$$

$0 < 1$ ✓

RESULT $\quad \sum_{n=1}^{\infty} \frac{10^n}{n!}$ is convergent.

> $\sum_{n=1}^{\infty} \frac{c^n}{n!}$ is convergent, for $c \ge 0$.

15. Basic series

Convergent

- $\Sigma \frac{1}{n^p}$ for $p \ge 2$ — Frame 5
- Σc^n for $0 \le c < 1$ — Geometric Series
- $\Sigma n^p c^n$ for $p > 0, 0 \le c < 1$ — Frame 13
- $\Sigma \frac{c^n}{n!}$ for $c \ge 0$ — Frame 14

Divergent

- $\Sigma \frac{1}{n^p}$ for $p \le 1$ — Frame 6

Exercise 2.3 Use the Ratio Test to determine whether the following series are convergent.

(a) $\sum_{n=1}^{\infty} \frac{n^3}{n!}$ (b) $\sum_{n=1}^{\infty} \frac{n^2 2^n}{n!}$ (c) $\sum_{n=1}^{\infty} \frac{(2n)!}{n^n}$

Hint: In part (c) you need to use the fact that $(1 + 1/n)^n \to e$ as $n \to \infty$.

When using the Ratio Test, you obtain a_{n+1} by replacing *each* instance of n by $n + 1$ in the formula for a_n.

Remarks

1. The second part of the Ratio Test states that

   if $\sum a_n$ has positive terms and $a_{n+1}/a_n \to l$ as $n \to \infty$, where $l > 1$, then $\sum a_n$ is divergent.

   We prove this later by deducing that $\{a_n\}$ is not a null sequence, and applying the Non-null Test. Thus you may be able to avoid using the Ratio Test by recognising that $\{a_n\}$ is not null. For example, we know that $\{n!/10^n\}$ is not a null sequence, so, by the Non-null Test,

   $$\sum_{n=1}^{\infty} \frac{n!}{10^n} \text{ is divergent.}$$

   See Subsection 2.2.

   Since $\{10^n/n!\}$ is a basic null sequence,
   $$n!/10^n \to \infty \quad \text{as } n \to \infty,$$
   by the Reciprocal Rule; see Unit AA2, Section 4.

2. The first part of the Comparison Test states that

   if $0 \le a_n \le b_n$, for $n = 1, 2, \ldots$, and $\sum_{n=1}^{\infty} b_n$ is convergent, then $\sum_{n=1}^{\infty} a_n$ is convergent.

   In fact, the proof enables us to deduce more than this, namely, that

   $$\sum_{n=1}^{\infty} a_n \le \sum_{n=1}^{\infty} b_n. \tag{2.1}$$

   See Subsection 2.2.

3. To apply the first part of the Comparison Test, it is sufficient to check that the inequality $0 \le a_n \le b_n$ holds *eventually*; that is,

   $$0 \le a_n \le b_n, \quad \text{for } n \ge N, \tag{2.2}$$

   where N is some positive integer. Similarly, to apply the second part of the Comparison Test, it is sufficient to check that

   $$0 \le b_n \le a_n, \quad \text{for } n \ge N,$$

   where N is some positive integer.

   However, if equation (2.2) holds with some $N > 1$, then we cannot deduce that equation (2.1) holds.

4. The Comparison Test tells us that any series of the form

   $$a_0 + \sum_{n=1}^{\infty} \frac{a_n}{10^n},$$

   where a_0 is a non-negative integer and a_n, $n = 1, 2, \ldots$, are digits, must be convergent. Moreover, the partial sums of this series are $s_n = a_0.a_1a_2\ldots a_n$, so its sum is $a_0.a_1a_2\ldots$. Thus this series provides an alternative interpretation of the infinite decimal $a_0.a_1a_2\ldots$.

   For example, by this reasoning, the series

   $$\frac{3}{10^1} + \frac{3}{10^2} + \frac{3}{10^3} + \cdots \quad \text{has sum } 0.333\ldots,$$

   which agrees with the result obtained by summing the geometric series with $a = \frac{3}{10}$ and $r = \frac{1}{10}$:

   $$\frac{a}{1-r} = \frac{\frac{3}{10}}{1 - \frac{1}{10}} = \frac{1}{3}.$$

   This approach gives another method of finding the rational which is represented by a given recurring decimal; see Unit AA1, Section 1.

2.2 Proofs

We now supply proofs of the tests given in the previous subsection.

If you are short of time, omit these proofs.

> **Comparison Test**
>
> (a) If
>
> $$0 \le a_n \le b_n, \quad \text{for } n = 1, 2, \ldots, \tag{2.3}$$
>
> and $\sum_{n=1}^{\infty} b_n$ is convergent, then $\sum_{n=1}^{\infty} a_n$ is convergent.
>
> (b) If
>
> $$0 \le b_n \le a_n, \quad \text{for } n = 1, 2, \ldots, \tag{2.4}$$
>
> and $\sum_{n=1}^{\infty} b_n$ is divergent, then $\sum_{n=1}^{\infty} a_n$ is divergent.

Proof

(a) We assume here that inequality (2.3) holds. Thus the nth partial sums

$$s_n = a_1 + a_2 + \cdots + a_n, \quad n = 1, 2, \ldots,$$

and

$$t_n = b_1 + b_2 + \cdots + b_n, \quad n = 1, 2, \ldots,$$

satisfy

$$s_n \le t_n, \quad \text{for } n = 1, 2, \ldots. \tag{2.5}$$

We also know that $\sum_{n=1}^{\infty} b_n$ is convergent, so the increasing sequence $\{t_n\}$ is convergent with limit t, say. Hence

$$s_n \le t_n \le t, \quad \text{for } n = 1, 2, \ldots,$$

so the increasing sequence $\{s_n\}$ is bounded above by t. By the Monotone Convergence Theorem, $\{s_n\}$ is also convergent, so $\sum_{n=1}^{\infty} a_n$ is a convergent series.

(b) We have to show that when inequality (2.4) holds:

$$\text{if } \sum_{n=1}^{\infty} a_n \text{ is convergent, then } \sum_{n=1}^{\infty} b_n \text{ is also convergent.}$$

This is a proof by contraposition.

This implication follows immediately from part (a). ■

Remark In the proof of part (a), inequality (2.5) implies that

$$\sum_{n=1}^{\infty} a_n \le \sum_{n=1}^{\infty} b_n,$$

by the Limit Inequality Rule.

See Unit AA2, Section 3.

Next we prove the Limit Comparison Test.

Limit Comparison Test Suppose that $\sum_{n=1}^{\infty} a_n$ and $\sum_{n=1}^{\infty} b_n$ have positive terms and that

$$\frac{a_n}{b_n} \to L \quad \text{as } n \to \infty, \tag{2.6}$$

where $L \neq 0$.

(a) If $\sum_{n=1}^{\infty} b_n$ is convergent, then $\sum_{n=1}^{\infty} a_n$ is convergent.

(b) If $\sum_{n=1}^{\infty} b_n$ is divergent, then $\sum_{n=1}^{\infty} a_n$ is divergent.

Proof

(a) We know that the sequence $\{a_n/b_n\}$ is convergent, so it must be bounded. Thus there is a constant K such that

See Unit AA2, Theorem 4.1.

$$\frac{a_n}{b_n} \leq K, \quad \text{for } n = 1, 2, \ldots,$$

so

$$a_n \leq K b_n, \quad \text{for } n = 1, 2, \ldots.$$

We also know that $\sum_{n=1}^{\infty} b_n$ is convergent, so $\sum_{n=1}^{\infty} K b_n$ is convergent, by the Multiple Rule. Hence, by the Comparison Test, $\sum_{n=1}^{\infty} a_n$ is convergent.

(b) We have to show that when limit (2.6) holds:

$$\text{if } \sum_{n=1}^{\infty} a_n \text{ is convergent, then } \sum_{n=1}^{\infty} b_n \text{ is also convergent.}$$

This is again a proof by contraposition.

This implication follows immediately from part (a) because

$$\frac{b_n}{a_n} \to \frac{1}{L} \quad \text{as } n \to \infty,$$

Remember that $L \neq 0$.

by the Quotient Rule for sequences. ■

Remark The assumption that $L \neq 0$ is not needed in the proof of part (a), but it is essential in the proof of part (b).

The final test given in the audio frames is the Ratio Test.

Ratio Test Suppose that $\sum_{n=1}^{\infty} a_n$ has positive terms and that

$$\frac{a_{n+1}}{a_n} \to l \text{ as } n \to \infty.$$

(a) If $0 \leq l < 1$, then $\sum_{n=1}^{\infty} a_n$ is convergent.

(b) If $l > 1$, then $\sum_{n=1}^{\infty} a_n$ is divergent.

Part (b) includes the case $\dfrac{a_{n+1}}{a_n} \to \infty$.

Proof

(a) We know that $0 \leq l < 1$, so we can choose $\varepsilon > 0$ such that

For example, take $\varepsilon = \frac{1}{2}(1 - l)$.

$$l + \varepsilon < 1.$$

Let $r = l + \varepsilon$. Since $r > l$, there is a positive integer N such that

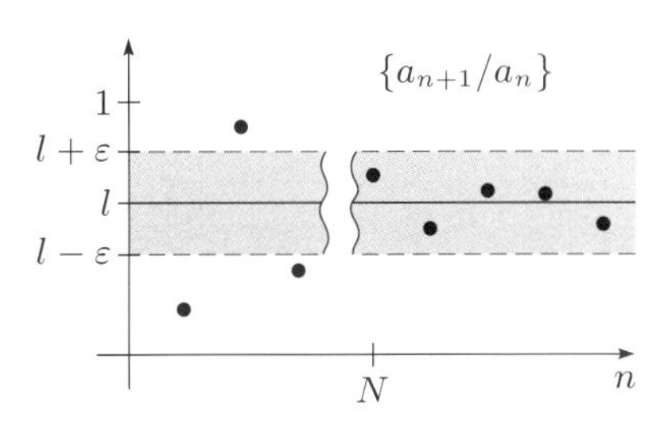

$$\frac{a_{n+1}}{a_n} \leq r, \quad \text{for all } n \geq N.$$

Thus, for $n \geq N$, we have

$$\frac{a_n}{a_N} = \left(\frac{a_n}{a_{n-1}}\right)\left(\frac{a_{n-1}}{a_{n-2}}\right)\cdots\left(\frac{a_{N+1}}{a_N}\right) \leq r^{n-N},$$

since each of the expressions in brackets is at most r. Hence

$$a_n \leq a_N r^{n-N}, \quad \text{for } n \geq N. \tag{2.7}$$

Now

$$\sum_{n=N}^{\infty} a_N r^{n-N} = a_N + a_N r + a_N r^2 + \cdots$$

is a geometric series with first term a_N and common ratio r. Since $0 < r < 1$, this series is convergent. Thus, by inequality (2.7) and the Comparison Test, $\sum_{n=1}^{\infty} a_n$ is also convergent, as required.

(b) Since

$$\frac{a_{n+1}}{a_n} \to \infty \quad \text{or} \quad \frac{a_{n+1}}{a_n} \to l,$$

where $l > 1$, there is a positive integer N such that

$$\frac{a_{n+1}}{a_n} \geq 1, \quad \text{for all } n \geq N.$$

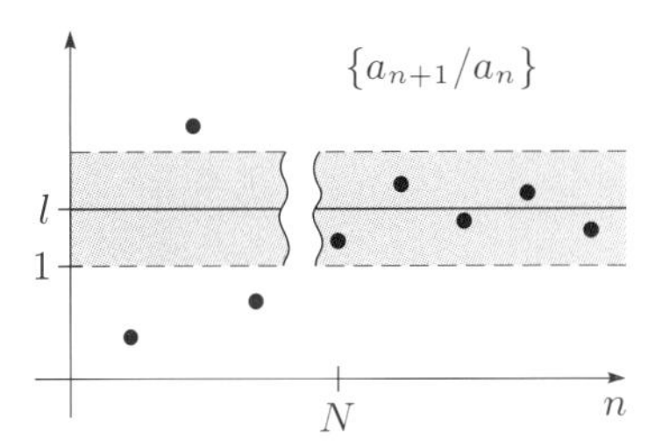

Thus, for $n \geq N$, we have

$$\frac{a_n}{a_N} = \left(\frac{a_n}{a_{n-1}}\right)\left(\frac{a_{n-1}}{a_{n-2}}\right)\cdots\left(\frac{a_{N+1}}{a_N}\right) \geq 1,$$

since each of the expressions in brackets is at least 1. Hence

$$a_n \geq a_N > 0, \quad \text{for } n \geq N,$$

so $\{a_n\}$ cannot be a null sequence. It follows, by the Non-null Test, that $\sum_{n=1}^{\infty} a_n$ is divergent. ■

We end this section by proving the convergence or divergence of the basic series given in the final frame of the audio section.

Basic series The following series are convergent:

(a) $\displaystyle\sum_{n=1}^{\infty} \frac{1}{n^p}$, for $p \geq 2$;

(b) $\displaystyle\sum_{n=1}^{\infty} c^n$, for $0 \leq c < 1$;

(c) $\displaystyle\sum_{n=1}^{\infty} n^p c^n$, for $p > 0$, $0 \leq c < 1$;

(d) $\displaystyle\sum_{n=1}^{\infty} \frac{c^n}{n!}$, for $c \geq 0$.

The following series is divergent:

(e) $\displaystyle\sum_{n=1}^{\infty} \frac{1}{n^p}$, for $0 < p \leq 1$.

In Unit AB3 we shall prove that $\displaystyle\sum_{n=1}^{\infty} \frac{1}{n^p}$ is convergent, for all $p > 1$.

Proof

(a) As indicated in Frame 5, this series is convergent by the Comparison Test, since if $p \geq 2$, then

$$\frac{1}{n^p} \leq \frac{1}{n^2}, \quad \text{for } n = 1, 2, \ldots.$$

(b) The series $\displaystyle\sum_{n=1}^{\infty} c^n$ is a geometric series with common ratio c, so it converges if $0 \leq c < 1$.

See page 9.

(c) Let

$$a_n = n^p c^n, \quad n = 1, 2, \ldots.$$

Then put $b = \sqrt{c}$ and express a_n as

$$a_n = n^p (b \times b)^n = (n^p b^n) b^n, \quad \text{for } n = 1, 2, \ldots. \tag{2.8}$$

Now $0 \leq b < 1$, so $\{n^p b^n\}$ is a basic null sequence. Hence, for some positive integer N, we have

$$n^p b^n < 1, \quad \text{for } n > N,$$

and thus, by equation (2.8),

$$a_n < b^n, \quad \text{for } n > N.$$

But $\sum b^n$ is a convergent geometric series, so $\sum a_n$ is convergent by the Comparison Test.

See remark 3 after the audio frames.

(d) Let

$$a_n = \frac{c^n}{n!}, \quad n = 1, 2, \ldots.$$

Then, for $c \neq 0$,

$$\frac{a_{n+1}}{a_n} = \frac{c^{n+1}}{(n+1)!} \Big/ \frac{c^n}{n!} = \left(\frac{c^{n+1}}{(n+1)!}\right)\left(\frac{n!}{c^n}\right) = \frac{c}{n+1}.$$

If $c = 0$, then the series is clearly convergent.

Thus

$$\frac{a_{n+1}}{a_n} \to 0 \quad \text{as } n \to \infty$$

and we deduce from the Ratio Test that

$$\sum_{n=1}^{\infty} \frac{c^n}{n!} \text{ is convergent.}$$

(e) As indicated in Frame 6, this series is divergent by the Comparison Test, since if $p \leq 1$, then

$$\frac{1}{n^p} \geq \frac{1}{n}, \quad \text{for } n = 1, 2, \ldots. \quad ■$$

Further exercises

Exercise 2.4 Determine whether or not the following series are convergent.

(a) $\displaystyle\sum_{n=1}^{\infty} \frac{\cos(1/n)}{2n^2+3}$ (b) $\displaystyle\sum_{n=1}^{\infty} \frac{n^2}{2n^3-n}$ (c) $\displaystyle\sum_{n=1}^{\infty} \frac{\sqrt{2n}}{4n^3+n+2}$

(d) $\displaystyle\sum_{n=1}^{\infty} \frac{(n+1)^5}{2^n}$ (e) $\displaystyle\sum_{n=1}^{\infty} \frac{n^2 3^n}{n!}$ (f) $\displaystyle\sum_{n=1}^{\infty} \frac{(n!)^2}{(2n)!}$

Exercise 2.5 Use the Ratio Test to show that:

(a) $\displaystyle\sum_{n=1}^{\infty} \frac{2^n n!}{n^n}$ converges; (b) $\displaystyle\sum_{n=1}^{\infty} \frac{3^n n!}{n^n}$ diverges.

3 Series with positive and negative terms

After working through this section, you should be able to:

(a) explain the term *absolutely convergent* and use the Absolute Convergence Test;

(b) use the Alternating Test;

(c) use the given general strategy for testing a series for convergence.

The study of series $\sum a_n$ with $a_n \geq 0$ for all values of n is relatively straightforward because the sequence $\{s_n\}$ of partial sums is increasing. Similarly, if $a_n \leq 0$ for all values of n, then $\{s_n\}$ is decreasing.

It is harder to determine the behaviour of a series with both positive and negative terms because $\{s_n\}$ is neither increasing nor decreasing. However, if the sequence $\{a_n\}$ contains only finitely many negative terms, then the sequence $\{s_n\}$ is *eventually* increasing, and we can apply the results of Section 2. Similarly, if $\{a_n\}$ contains only finitely many positive terms, then the sequence $\{s_n\}$ is *eventually* decreasing, and we can again apply the results of Section 2, after making a sign change.

For example, the convergence of

$$1+2+3-\frac{1}{4^2}-\frac{1}{5^2}-\frac{1}{6^2}-\cdots$$

follows from that of $\sum_{n=1}^{\infty} \frac{1}{n^2}$, by the Multiple Rule with $\lambda = -1$.

In this section we look at series such as

$$1-\frac{1}{2}+\frac{1}{3}-\frac{1}{4}+\frac{1}{5}-\frac{1}{6}+\cdots$$

and

$$1+\frac{1}{2^2}-\frac{1}{3^2}+\frac{1}{4^2}+\frac{1}{5^2}-\frac{1}{6^2}+\cdots,$$

which contain infinitely many terms of either sign. For such series the partial sums increase and decrease infinitely often (so the snake discussed on page 7 grows and shrinks infinitely often!). We give two methods which can often be used to prove that such series are convergent.

3.1 Absolute convergence

Suppose that we want to determine the behaviour of the infinite series

$$\sum_{n=1}^{\infty} \frac{(-1)^{n+1}}{n^2} = 1 - \frac{1}{2^2} + \frac{1}{3^2} - \frac{1}{4^2} + \frac{1}{5^2} - \frac{1}{6^2} + \cdots. \qquad (3.1)$$

We know that the series

$$\sum_{n=1}^{\infty} \frac{1}{n^2} = 1 + \frac{1}{2^2} + \frac{1}{3^2} + \frac{1}{4^2} + \frac{1}{5^2} + \frac{1}{6^2} + \cdots \qquad (3.2)$$

Series (3.2) is a basic convergent series.

is convergent. Does this imply that series (3.1) is also convergent? In fact it does, as we now prove.

Consider the two related series

$$1 + 0 + \frac{1}{3^2} + 0 + \frac{1}{5^2} + 0 + \cdots$$

and

$$0 + \frac{1}{2^2} + 0 + \frac{1}{4^2} + 0 + \frac{1}{6^2} + \cdots.$$

Each of these series contains only non-negative terms and is dominated by series (3.2), so they are both convergent, by the Comparison Test. Applying the Sum Rule, and the Multiple Rule with $\lambda = -1$, we deduce that the series

$$1 - \frac{1}{2^2} + \frac{1}{3^2} - \frac{1}{4^2} + \frac{1}{5^2} - \frac{1}{6^2} + \cdots \text{ is convergent.}$$

The type of argument just given is the basis for a concept called *absolute convergence*, which we now define.

> **Definition** The series $\sum_{n=1}^{\infty} a_n$ is **absolutely convergent** if $\sum_{n=1}^{\infty} |a_n|$ is convergent.

If the terms a_n are all non-negative, then *absolute convergence* and *convergence* have the same meaning.

For example, series (3.1) is absolutely convergent because the series $\sum_{n=1}^{\infty} 1/n^2$ is convergent.

However, the series

$$\sum_{n=1}^{\infty} \frac{(-1)^{n+1}}{n} = 1 - \frac{1}{2} + \frac{1}{3} - \frac{1}{4} + \frac{1}{5} - \frac{1}{6} + \cdots \qquad (3.3)$$

is not absolutely convergent because the series $\sum_{n=1}^{\infty} 1/n$ is divergent.

As the name suggests, every absolutely convergent series is convergent.

Absolute Convergence Test If $\sum_{n=1}^{\infty} a_n$ is absolutely convergent, then $\sum_{n=1}^{\infty} a_n$ is convergent.

The proofs of the results in this section are in Subsection 3.4.

Because $\sum_{n=1}^{\infty} 1/n^2$ is convergent, it follows from the Absolute Convergence Test that series (3.1) is convergent, as we have already seen. Indeed, however we distribute the plus and minus signs amongst the terms of $\{1/n^2\}$, the resulting series is convergent.

However, the Absolute Convergence Test tells us nothing about the behaviour of series (3.3), nor about the series

We decide later, using other methods, whether series (3.3) and (3.4) are convergent.

$$1+\frac{1}{2}-\frac{1}{3}+\frac{1}{4}+\frac{1}{5}-\frac{1}{6}+\frac{1}{7}+\frac{1}{8}-\frac{1}{9}+\cdots. \qquad (3.4)$$

The series $\sum_{n=1}^{\infty} 1/n$ is divergent, so these two series are not absolutely convergent.

Example 3.1 Prove that the following series are convergent.

(a) $\displaystyle\sum_{n=1}^{\infty} \frac{(-1)^{n+1}}{n^3}$ (b) $\displaystyle\sum_{n=1}^{\infty} \frac{\cos n}{2^n}$

Solution

(a) Let

$$a_n = \frac{(-1)^{n+1}}{n^3}, \quad n = 1, 2, \ldots; \quad \text{then} \quad |a_n| = \frac{1}{n^3}, \quad \text{for } n = 1, 2, \ldots.$$

We know that

$$\sum_{n=1}^{\infty} \frac{1}{n^3} \text{ is convergent,} \quad \text{so} \quad \sum_{n=1}^{\infty} \frac{(-1)^{n+1}}{n^3} \text{ is absolutely convergent.}$$

Hence, by the Absolute Convergence Test,

$$\sum_{n=1}^{\infty} \frac{(-1)^{n+1}}{n^3} \text{ is convergent.}$$

(b) Let

$$a_n = \frac{\cos n}{2^n}, \quad n = 1, 2, \ldots; \quad \text{then} \quad |a_n| \leq \frac{1}{2^n}, \quad \text{for } n = 1, 2, \ldots,$$

because $|\cos n| \leq 1$, for $n = 1, 2, \ldots$.

Thus $\sum_{n=1}^{\infty} |a_n|$ is convergent, by the Comparison Test, since $\sum_{n=1}^{\infty} \frac{1}{2^n}$ is a convergent geometric series, so $\sum_{n=1}^{\infty} a_n$ is absolutely convergent. Hence, by the Absolute Convergence Test,

$$\sum_{n=1}^{\infty} \frac{\cos n}{2^n} \text{ is convergent.} \quad \blacksquare$$

Exercise 3.1 Prove that the following series are convergent.

(a) $\displaystyle\sum_{n=1}^{\infty} \frac{(-1)^{n+1}n}{n^3+1}$ (b) $1 + \frac{1}{2} - \frac{1}{4} + \frac{1}{8} + \frac{1}{16} - \frac{1}{32} + \cdots$

The Absolute Convergence Test states that if the series $\sum |a_n|$ is convergent, then $\sum a_n$ is also convergent. The following result relates the sums of these two convergent series.

> **Triangle Inequality (infinite form)** If $\displaystyle\sum_{n=1}^{\infty} a_n$ is absolutely convergent, then
>
> $$\left|\sum_{n=1}^{\infty} a_n\right| \leq \sum_{n=1}^{\infty} |a_n|.$$

This generalises the inequality

$$\left|\sum_{k=1}^{n} a_k\right| \leq \sum_{k=1}^{n} |a_k|;$$

see Unit AA1, Subsection 3.1, after the Triangle Inequality.

Exercise 3.2 Show that the series

$$\frac{1}{2} - \frac{1}{4} - \frac{1}{8} + \frac{1}{16} - \frac{1}{32} - \frac{1}{64} + \cdots$$

is convergent, and that its sum lies in $[-1, 1]$. (You do not need to find the sum of the series.)

3.2 Alternating Test

Suppose that we want to determine the behaviour of the following infinite series, in which the terms have alternating signs:

$$\sum_{n=1}^{\infty} \frac{(-1)^{n+1}}{n} = 1 - \frac{1}{2} + \frac{1}{3} - \frac{1}{4} + \frac{1}{5} - \frac{1}{6} + \cdots.$$

The Absolute Convergence Test does not help us with this series because $\displaystyle\sum_{n=1}^{\infty} 1/n$ is divergent.

In fact, the above series is convergent. To see why, we first calculate some of the partial sums and plot them on a sequence diagram:

$$s_1 = 1,$$

$$s_2 = 1 - \frac{1}{2} = 0.5,$$

$$s_3 = 1 - \frac{1}{2} + \frac{1}{3} = 0.8\overline{3},$$

$$s_4 = 1 - \frac{1}{2} + \frac{1}{3} - \frac{1}{4} = 0.58\overline{3},$$

$$s_5 = 1 - \frac{1}{2} + \frac{1}{3} - \frac{1}{4} + \frac{1}{5} = 0.78\overline{3},$$

$$s_6 = 1 - \frac{1}{2} + \frac{1}{3} - \frac{1}{4} + \frac{1}{5} - \frac{1}{6} = 0.61\overline{6}.$$

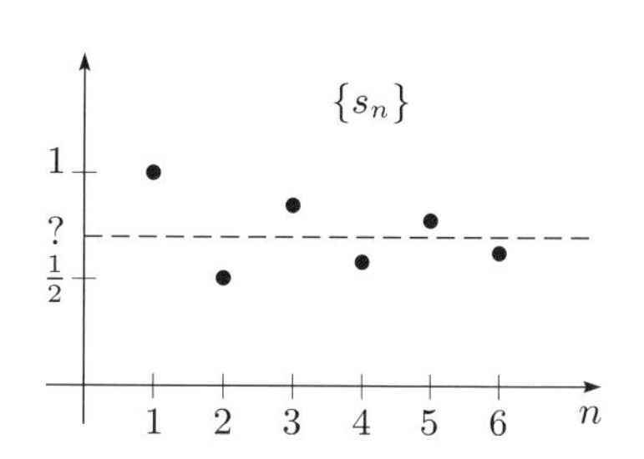

The sequence diagram suggests that the odd subsequence $\{s_{2k-1}\}$ is decreasing:

$$s_1 \geq s_3 \geq s_5 \geq \cdots \geq s_{2k-1} \geq \cdots,$$

and that the even subsequence $\{s_{2k}\}$ is increasing:

$$s_2 \leq s_4 \leq s_6 \leq \cdots \leq s_{2k} \leq \cdots.$$

Subsequences were introduced in Unit AA2, Subsection 4.4.

Also, the terms of $\{s_{2k-1}\}$ all exceed the terms of $\{s_{2k}\}$, and both subsequences appear to converge to a common limit s, which lies between the odd and even partial sums.

To prove this, we write the even partial sum s_{2k} as

$$s_{2k} = \left(1 - \frac{1}{2}\right) + \left(\frac{1}{3} - \frac{1}{4}\right) + \cdots + \left(\frac{1}{2k-1} - \frac{1}{2k}\right).$$

All the expressions in brackets are positive, so the subsequence $\{s_{2k}\}$ is increasing.

We can also write

$$s_{2k} = 1 - \left(\frac{1}{2} - \frac{1}{3}\right) - \left(\frac{1}{4} - \frac{1}{5}\right) - \cdots - \left(\frac{1}{2k-2} - \frac{1}{2k-1}\right) - \frac{1}{2k}.$$

Again, all the expressions in brackets are positive, so $\{s_{2k}\}$ is bounded above by 1.

Hence, by the Monotone Convergence Theorem,

$$\lim_{k\to\infty} s_{2k} = s,$$

for some s. Since

$$s_{2k} = s_{2k-1} - \frac{1}{2k}$$

and $\{1/(2k)\}$ is null, we deduce that

$$\lim_{k\to\infty} s_{2k-1} = \lim_{k\to\infty} \left(s_{2k} + \frac{1}{2k}\right) = s,$$

by the Sum Rule for sequences. Thus the odd and even subsequences of $\{s_n\}$ both tend to the same limit s, so $s_n \to s$ as $n \to \infty$. Hence, the series

See Unit AA2, Theorem 4.3.

$$1 - \frac{1}{2} + \frac{1}{3} - \frac{1}{4} + \frac{1}{5} - \frac{1}{6} + \cdots \text{ is convergent, with sum } s.$$

In fact, $s = \log_e 2 \simeq 0.69$.

The same method can be used to prove the following general result.

Alternating Test Let

$$a_n = (-1)^{n+1} b_n, \quad n = 1, 2, \ldots,$$

where $\{b_n\}$ is a decreasing null sequence with positive terms. Then

$$\sum_{n=1}^{\infty} a_n = b_1 - b_2 + b_3 - b_4 + \cdots \text{ is convergent.}$$

This test is also called the Leibniz Test.

When you apply the Alternating Test, there are a number of conditions which must be checked. We now describe these in the form of a strategy.

Strategy 3.1 To prove that $\sum_{n=1}^{\infty} a_n$ is convergent using the Alternating Test, check that

$$a_n = (-1)^{n+1} b_n, \quad n = 1, 2, \ldots,$$

where

1. $b_n \geq 0$, for $n = 1, 2, \ldots$;
2. $\{b_n\}$ is a null sequence;
3. $\{b_n\}$ is decreasing.

To show that $\{b_n\}$ is null, use the techniques introduced in Unit AA2 and the basic null sequences listed on page 7 of this unit.

Here are some examples of the use of Strategy 3.1.

Example 3.2 Prove that the following series are convergent.

(a) $\sum_{n=1}^{\infty} \frac{(-1)^{n+1}}{\sqrt{n}}$ (b) $\sum_{n=1}^{\infty} \frac{(-1)^{n+1}}{n^4}$

Solution

(a) The sequence $\{(-1)^{n+1}/\sqrt{n}\}$ is of the form $\{(-1)^{n+1} b_n\}$, where

$$b_n = 1/\sqrt{n}, \quad n = 1, 2, \ldots.$$

Now

1. $1/\sqrt{n} \geq 0$, for $n = 1, 2, \ldots$;
2. $\{1/\sqrt{n}\}$ is a basic null sequence;
3. $\{1/\sqrt{n}\}$ is decreasing (because $\{\sqrt{n}\}$ is increasing).

Hence, by the Alternating Test,

$$\sum_{n=1}^{\infty} \frac{(-1)^{n+1}}{\sqrt{n}} \text{ is convergent.}$$

(b) The sequence $\{(-1)^{n+1}/n^4\}$ is of the form $\{(-1)^{n+1} b_n\}$, where

$$b_n = 1/n^4, \quad n = 1, 2, \ldots.$$

Now

1. $1/n^4 \geq 0$, for $n = 1, 2, \ldots$;
2. $\{1/n^4\}$ is a basic null sequence;
3. $\{1/n^4\}$ is decreasing (because $\{n^4\}$ is increasing).

Hence, by the Alternating Test,

$$\sum_{n=1}^{\infty} \frac{(-1)^{n+1}}{n^4} \text{ is convergent.} \quad \blacksquare$$

Alternatively, you can show that this series is convergent by using the Absolute Convergence Test, as in Example 3.1(a).

Exercise 3.3 Determine which of the following series are convergent.

(a) $\sum_{n=1}^{\infty} \frac{(-1)^{n+1}}{n^{1/3}}$ (b) $\sum_{n=1}^{\infty} \frac{(-1)^{n+1}}{n + n^{1/2}}$ (c) $\sum_{n=1}^{\infty} (-1)^{n+1} \frac{n}{n+2}$

3.3 General strategy

We now give a strategy for applying the tests for convergence (or divergence) of a series. For each test, the strategy briefly indicates the circumstances under which that test can be used.

Strategy 3.2 General strategy for convergence or divergence

1. If you think that the sequence of terms $\{a_n\}$ is non-null, then try the **Non-null Test**.
2. If $\sum a_n$ has non-negative terms, then try one of these tests.
   (a) **Basic series** Is $\sum a_n$ a basic series, or a combination of these?
   (b) **Comparison Test** Is $a_n \leq b_n$, where $\sum b_n$ is convergent, or $a_n \geq b_n \geq 0$, where $\sum b_n$ is divergent?
   (c) **Limit Comparison Test** Does a_n behave like b_n (that is, $a_n/b_n \to L \neq 0$), where $\sum b_n$ is a basic series?
   (d) **Ratio Test** Does $a_{n+1}/a_n \to l \neq 1$?
3. If $\sum a_n$ has positive and negative terms, then try one of these tests.
   (a) **Absolute Convergence Test** Is $\sum |a_n|$ convergent? (Use step 2.)
   (b) **Alternating Test** Is $a_n = (-1)^{n+1} b_n$, where $\{b_n\}$ is non-negative, null and decreasing?

The following suggestions may also be helpful.

If a_n is positive and includes $n!$ or c^n, then consider the Ratio Test.

If a_n is positive and has dominant term n^p, then consider the (Limit) Comparison Test.

If a_n includes a sine or cosine term, then use the fact that this term is bounded and consider the Comparison Test and the Absolute Convergence Test.

Remarks

1. When applying these tests, you do not need to try to prove that the sequence $\{a_n\}$ is null (except in the case of the Alternating Test).
2. If none of steps 1–3 gives a result, then you can try using first principles by working directly with the sequence $\{s_n\}$ of partial sums.

See Exercise 3.6, for example.

Exercise 3.4 Determine which of the following series are convergent.

(a) $\displaystyle\sum_{n=1}^{\infty} \frac{5n+2^n}{3^n}$ (b) $\displaystyle\sum_{n=1}^{\infty} \frac{3}{2n^3-1}$ (c) $\displaystyle\sum_{n=1}^{\infty} \frac{(-1)^{n+1}}{n\log_e(n+1)}$

(d) $\displaystyle\sum_{n=1}^{\infty} \frac{(-1)^{n+1}n^2}{n^2+1}$ (e) $\displaystyle\sum_{n=1}^{\infty} \frac{(-1)^{n+1}n}{n^3+5}$ (f) $\displaystyle\sum_{n=1}^{\infty} \frac{2^n}{n^6}$

3.4 Proofs

We now supply the proofs omitted earlier in the section.

If you are short of time, omit these proofs.

Absolute Convergence Test If $\displaystyle\sum_{n=1}^{\infty} a_n$ is absolutely convergent, then $\displaystyle\sum_{n=1}^{\infty} a_n$ is convergent.

Proof We know that $\sum_{n=1}^{\infty} |a_n|$ is convergent, and we want to prove that $\sum_{n=1}^{\infty} a_n$ is convergent.

To do this, we define two new series $\sum_{n=1}^{\infty} b_n$ and $\sum_{n=1}^{\infty} c_n$, where

$$b_n = \begin{cases} a_n, & \text{if } a_n \geq 0, \\ 0, & \text{if } a_n < 0, \end{cases} \qquad c_n = \begin{cases} 0, & \text{if } a_n \geq 0, \\ -a_n, & \text{if } a_n < 0. \end{cases}$$

Both the series $\sum_{n=1}^{\infty} b_n$ and $\sum_{n=1}^{\infty} c_n$ have non-negative terms, and

$$b_n \leq |a_n|, \quad \text{for } n = 1, 2, \ldots, \tag{3.5}$$

and

$$c_n \leq |a_n|, \quad \text{for } n = 1, 2, \ldots. \tag{3.6}$$

Since $\sum_{n=1}^{\infty} |a_n|$ is convergent, we deduce that $\sum_{n=1}^{\infty} b_n$ and $\sum_{n=1}^{\infty} c_n$ are convergent, by the Comparison Test. Thus

$$\sum_{n=1}^{\infty} a_n = \sum_{n=1}^{\infty} (b_n - c_n) = \sum_{n=1}^{\infty} b_n - \sum_{n=1}^{\infty} c_n \tag{3.7}$$

is convergent, by the Combination Rules. ■

Triangle Inequality (infinite form) If $\sum_{n=1}^{\infty} a_n$ is absolutely convergent, then

$$\left| \sum_{n=1}^{\infty} a_n \right| \leq \sum_{n=1}^{\infty} |a_n|.$$

Proof We use the series $\sum_{n=1}^{\infty} b_n$ and $\sum_{n=1}^{\infty} c_n$, introduced in the proof of the Absolute Convergence Test. From equation (3.7) we obtain

See remark 2 after the audio frames.

$$-\sum_{n=1}^{\infty} c_n \leq \sum_{n=1}^{\infty} a_n \leq \sum_{n=1}^{\infty} b_n.$$

Thus, by inequalities (3.5) and (3.6), we deduce that

$$-\sum_{n=1}^{\infty} |a_n| \leq \sum_{n=1}^{\infty} a_n \leq \sum_{n=1}^{\infty} |a_n|,$$

which gives the required inequality

$$\left| \sum_{n=1}^{\infty} a_n \right| \leq \sum_{n=1}^{\infty} |a_n|. \quad ■$$

Alternating Test Let

$$a_n = (-1)^{n+1} b_n, \quad n = 1, 2, \ldots,$$

where $\{b_n\}$ is a decreasing null sequence with positive terms. Then

$$\sum_{n=1}^{\infty} a_n = b_1 - b_2 + b_3 - b_4 + \cdots \text{ is convergent.}$$

Proof We can write any even partial sum s_{2k} as

$$s_{2k} = (b_1 - b_2) + (b_3 - b_4) + \cdots + (b_{2k-1} - b_{2k}).$$

Since $\{b_n\}$ is decreasing, all the expressions in brackets are non-negative, so the even subsequence $\{s_{2k}\}$ is increasing.

We can also write

$$s_{2k} = b_1 - (b_2 - b_3) - (b_4 - b_5) - \cdots - (b_{2k-2} - b_{2k-1}) - b_{2k}.$$

Again, all the expressions in brackets are non-negative, so $\{s_{2k}\}$ is bounded above by b_1.

Hence, by the Monotone Convergence Theorem,

$$\lim_{k\to\infty} s_{2k} = s,$$

for some s. Since

$$s_{2k} = s_{2k-1} - b_{2k}$$

and $\{b_n\}$ is null, we deduce that

$$\lim_{k\to\infty} s_{2k-1} = \lim_{k\to\infty} (s_{2k} + b_{2k}) = s,$$

by the Sum Rule for sequences. Thus the odd and even subsequences of $\{s_n\}$ both tend to the same limit s. Hence $\{s_n\}$ tends to s, so

See Unit AA2, Theorem 4.3.

$$\sum_{n=1}^{\infty} a_n \text{ is convergent, with sum } s. \quad \blacksquare$$

Further exercises

Exercise 3.5 Determine which of the following series are convergent.

(a) $\displaystyle\sum_{n=1}^{\infty} \frac{(-1)^{n+1}}{1+\sqrt{n}}$ (b) $\displaystyle\sum_{n=1}^{\infty} \frac{\sin n}{n^2}$; (c) $\displaystyle\sum_{n=1}^{\infty} \frac{(-1)^{n+1} n!}{n^4+3}$

(d) $\displaystyle\sum_{n=1}^{\infty} \frac{n+2^n}{3^n+5}$ (e) $\displaystyle\sum_{n=1}^{\infty} \frac{(-1)^{n+1} n}{n^2+2}$

Exercise 3.6 (Harder) Prove that

$$\frac{1}{3n-2} + \frac{1}{3n-1} - \frac{1}{3n} > \frac{1}{3n}, \quad \text{for } n = 1, 2, \ldots,$$

and deduce that the series

$$1 + \frac{1}{2} - \frac{1}{3} + \frac{1}{4} + \frac{1}{5} - \frac{1}{6} + \cdots \text{ is divergent.}$$

4 Exponential function

After working through this section, you should be able to:

(a) appreciate that there are two equivalent definitions of e^x;

(b) understand how the series definition of e^x enables us to prove that e is irrational, and that $e^{x+y} = e^x e^y$.

4.1 Definition of e^x

In Unit AA2, Section 5, we defined $e = 2.718\,28\ldots$ to be the limit

Ideally, you should watch the video programme before studying this section.

$$e = \lim_{n\to\infty} \left(1 + \frac{1}{n}\right)^n. \tag{4.1}$$

We also stated that if x is rational, then

$$e^x = \lim_{n\to\infty} \left(1 + \frac{x}{n}\right)^n, \tag{4.2}$$

and we verified this equation in some simple cases. We showed that the limit in equation (4.2) exists when x is positive, and we used equation (4.2) to *define* e^x when x is positive and irrational.

In the video programme for this unit, we plotted the partial sum functions

$$s_0(x) = 1, \quad s_1(x) = 1 + x, \quad s_2(x) = 1 + x + \frac{x^2}{2!}, \quad \ldots \quad ,$$

of the following infinite series of powers of x:

$$\sum_{n=0}^{\infty} \frac{x^n}{n!} = 1 + x + \frac{x^2}{2!} + \frac{x^3}{3!} + \cdots. \tag{4.3}$$

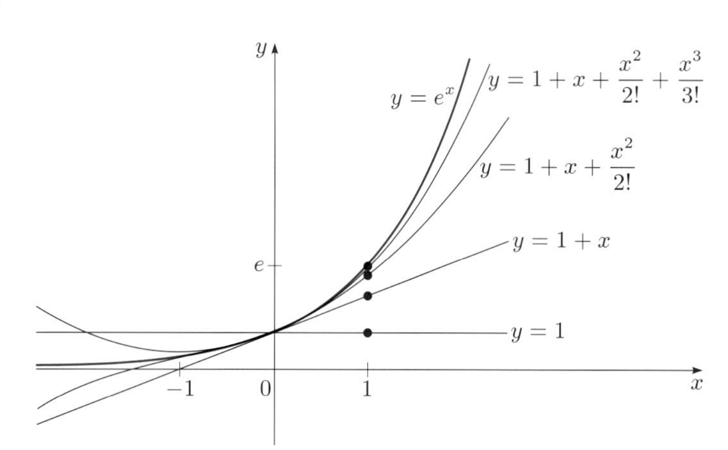

We know that series (4.3) is convergent for all real numbers x. Therefore the sum of the series depends on x, so it defines a function of x. In the video programme we found that this sum function appears to be e^x. In particular, when $x = 1$, the sum of the series

For $x \geq 0$, series (4.3) is a basic convergent series of type (d); see Subsection 2.2.

$$\sum_{n=0}^{\infty} \frac{1}{n!} = 1 + 1 + \frac{1}{2!} + \frac{1}{3!} + \cdots$$

is approximately $2.718\ldots$.

It can be proved that series (4.3) does converge to e^x, for all $x \in \mathbb{R}$, and we state this result, for $x > 0$, as our next theorem.

Theorem 4.1 If $x > 0$, then

$$\sum_{n=0}^{\infty} \frac{x^n}{n!} = \lim_{n\to\infty} \left(1 + \frac{x}{n}\right)^n = e^x.$$

Proof We give the proof only for the case $x = 1$; the general case is similar. We have to show that

If you are short of time, omit this proof.

$$\sum_{n=0}^{\infty} \frac{1}{n!} = \lim_{n\to\infty} \left(1 + \frac{1}{n}\right)^n = e.$$

The nth partial sum of this convergent series is

$$s_n = 1 + 1 + \frac{1}{2!} + \frac{1}{3!} + \cdots + \frac{1}{n!}, \quad \text{so} \quad \lim_{n\to\infty} s_n = \sum_{n=0}^{\infty} \frac{1}{n!}.$$

By the Binomial Theorem, we have

$$\left(1 + \frac{1}{n}\right)^n = 1 + n\left(\frac{1}{n}\right) + \frac{n(n-1)}{2!}\left(\frac{1}{n}\right)^2 + \cdots + \left(\frac{1}{n}\right)^n. \tag{4.4}$$

The general term in this expansion is of the form

$$\begin{aligned} \binom{n}{k}\left(\frac{1}{n}\right)^k &= \frac{n(n-1)(n-2)\ldots(n-k+1)}{k!}\left(\frac{1}{n}\right)^k \\ &= \frac{1}{k!}\left(1 - \frac{1}{n}\right)\left(1 - \frac{2}{n}\right)\ldots\left(1 - \frac{k-1}{n}\right), \end{aligned} \tag{4.5}$$

where $0 \le k \le n$. This last product is at most $1/k!$, since each expression in brackets is at most 1, so

$$\left(1 + \frac{1}{n}\right)^n \le 1 + 1 + \frac{1}{2!} + \cdots + \frac{1}{n!} = s_n,$$

by equation (4.4). Thus, by the Limit Inequality Rule for sequences,

$$e = \lim_{n\to\infty} \left(1 + \frac{1}{n}\right)^n \le \lim_{n\to\infty} s_n. \tag{4.6}$$

On the other hand, if $0 \le m \le n$, then (by equations (4.4) and (4.5))

$$\begin{aligned} &\left(1 + \frac{1}{n}\right)^n \\ &\ge 1 + 1 + \frac{1}{2!}\left(1 - \frac{1}{n}\right) + \cdots + \frac{1}{m!}\left(1 - \frac{1}{n}\right)\left(1 - \frac{2}{n}\right)\ldots\left(1 - \frac{m-1}{n}\right). \end{aligned}$$

Keeping m fixed and taking limits as $n \to \infty$, we obtain

$$e \ge 1 + 1 + \frac{1}{2!} + \cdots + \frac{1}{m!} = s_m,$$

by the Limit Inequality Rule. Using this rule once more, we obtain

$$\lim_{m\to\infty} s_m \le e. \tag{4.7}$$

Combining inequalities (4.6) and (4.7) gives $\sum_{n=0}^{\infty} \frac{1}{n!} = e$. ■

By Theorem 4.1, we can use the series $\sum_{n=0}^{\infty} x^n/n!$ as an alternative definition of e^x for all $x \geq 0$. We then define e^x for $x < 0$ as the reciprocal of e^{-x}.

For example,

$$e^{-\pi} = (e^{\pi})^{-1} = 1/e^{\pi}.$$

Definition

For $x \geq 0$,

$$e^x = \lim_{n\to\infty}\left(1+\frac{x}{n}\right)^n = \sum_{n=0}^{\infty}\frac{x^n}{n!}.$$

For $x < 0$,

$$e^x = (e^{-x})^{-1}.$$

The exponential function $x \longmapsto e^x$ is often called **exp**. Thus

$$\begin{aligned}\exp:\mathbb{R} &\longrightarrow \mathbb{R}\\ x &\longmapsto e^x.\end{aligned}$$

Remark The equations

$$e^x = \lim_{n\to\infty}\left(1+\frac{x}{n}\right)^n = \sum_{n=0}^{\infty}\frac{x^n}{n!}$$

are also true if x is negative, but we shall not prove this here.

The fact that

$$e^x = \sum_{n=0}^{\infty}\frac{x^n}{n!}, \quad \text{for } x < 0,$$

is proved in Unit AB4.

4.2 Calculating e

The representation of e by the infinite series

$$e = 1 + 1 + \frac{1}{2!} + \frac{1}{3!} + \cdots = \sum_{n=0}^{\infty}\frac{1}{n!}$$

provides a more efficient method of calculating approximate values for e than the equation $e = \lim_{n\to\infty}\left(1+\frac{1}{n}\right)^n$. This is illustrated by the following table of approximate values.

n	1	2	3	4	5
$(1+1/n)^n$	2	2.25	2.37	2.44	2.49
$\sum_{k=0}^{n}\frac{1}{k!}$	2	2.50	2.67	2.71	2.717

In fact,

$$e = 2.718\,281\,828\,45\ldots.$$

We now estimate how quickly the sequence of partial sums

$$s_n = 1 + 1 + \frac{1}{2!} + \cdots + \frac{1}{n!}, \quad n = 1, 2, \ldots,$$

converges to e. The difference between e and s_n is given by

$$\begin{aligned}e - s_n &= \frac{1}{(n+1)!} + \frac{1}{(n+2)!} + \frac{1}{(n+3)!} + \cdots\\ &= \frac{1}{(n+1)!}\left(1 + \frac{1}{n+2} + \frac{1}{(n+2)(n+3)} + \cdots\right)\\ &< \frac{1}{(n+1)!}\left(1 + \frac{1}{n+1} + \frac{1}{(n+1)^2} + \cdots\right).\end{aligned}$$

Here we replace each term by a larger term, so the new sum is greater than the previous one.

In the line above, the expression in large brackets is a geometric series with first term 1 and common ratio $1/(n+1)$, so its sum is

$$\frac{1}{1-1/(n+1)} = \frac{n+1}{n}.$$

Hence

$$0 < e - s_n < \frac{1}{(n+1)!} \times \frac{n+1}{n} = \frac{1}{n!} \times \frac{1}{n}, \quad \text{for } n = 1, 2, \ldots. \qquad (4.8)$$

For example,

$$0 < e - s_5 < \frac{1}{5!} \times \frac{1}{5} = 0.001\overline{6}.$$

Thus the difference between e and s_n is extremely small when n is large.

Inequality (4.8) can also be used to show that e is irrational.

> **Theorem 4.2** The number e is irrational.

Proof Suppose that $e = m/n$, where m and n are positive integers. Then, by inequality (4.8), for this particular integer n we have

This is a proof by contradiction.

$$0 < e - s_n < \frac{1}{n!} \times \frac{1}{n},$$

so

$$0 < n!(e - s_n) < \frac{1}{n}.$$

Since $e = m/n$, we have

$$0 < n!\left(\frac{m}{n} - \left(1 + 1 + \frac{1}{2!} + \cdots + \frac{1}{n!}\right)\right) < \frac{1}{n}.$$

But the middle expression in this pair of inequalities is an integer, as you can check by multiplying it out, so we have found an integer which lies strictly between 0 and 1. This is impossible, so e is not rational. ■

4.3 A fundamental property of e^x

We complete this section by showing that the function $f(x) = e^x$ satisfies one of the Exponent Laws.

We stated the Exponent Laws in Unit AA1, Section 5.

> **Theorem 4.3** For any real numbers x and y, we have $e^{x+y} = e^x e^y$.

Proof First we prove the special case where x and y are both positive.

If you are short of time, omit this proof.

The following table contains some of the terms which occur when we multiply together partial sums of the power series for e^x and e^y.

We have

$$e^x = 1 + x + \frac{x^2}{2!} + \frac{x^3}{3!} + \cdots$$

and

$$e^y = 1 + y + \frac{y^2}{2!} + \frac{y^2}{3!} + \cdots.$$

	1	y	$\frac{y^2}{2!}$	$\frac{y^3}{3!}$	$\cdots$
1	1	y	$\frac{y^2}{2!}$	$\frac{y^3}{3!}$	$\cdots$
x	x	xy	$\frac{xy^2}{2!}$	$\frac{xy^3}{3!}$	$\cdots$
$\frac{x^2}{2!}$	$\frac{x^2}{2!}$	$\frac{x^2y}{2!}$	$\frac{x^2y^2}{2!2!}$	$\frac{x^2y^3}{2!3!}$	$\cdots$
$\frac{x^3}{3!}$	$\frac{x^3}{3!}$	$\frac{x^3y}{3!}$	$\frac{x^3y^2}{3!2!}$	$\frac{x^3y^3}{3!3!}$	$\cdots$
$\vdots$	$\vdots$	$\vdots$	$\vdots$	$\vdots$	

Adding the terms on the 'lower left to upper right' diagonals of the table gives:

$$1 \qquad \text{(first diagonal)}$$

$$x+y \qquad \text{(second diagonal)}$$

$$\frac{x^2}{2!}+xy+\frac{y^2}{2!}=\frac{(x+y)^2}{2!} \qquad \text{(third diagonal)}$$

$$\frac{x^3}{3!}+\frac{x^2y}{2!}+\frac{xy^2}{2!}+\frac{y^3}{3!}=\frac{(x+y)^3}{3!} \qquad \text{(fourth diagonal)}$$

$$\vdots \qquad\qquad \vdots$$

$$\frac{x^n}{n!}+\frac{x^{n-1}y}{(n-1)!}+\cdots+\frac{xy^{n-1}}{(n-1)!}+\frac{y^n}{n!}=\frac{(x+y)^n}{n!} \qquad \text{((}n+1\text{)th diagonal).}$$

For any positive integer n, the product

$$\left(\sum_{k=0}^{n}\frac{x^k}{k!}\right)\left(\sum_{k=0}^{n}\frac{y^k}{k!}\right)=\left(1+x+\cdots+\frac{x^n}{n!}\right)\left(1+y+\cdots+\frac{y^n}{n!}\right)$$

includes *all* the terms in the first $n+1$ diagonals of the table; moreover, all the terms of the product lie in the first $2n+1$ diagonals.

Since x and y are non-negative, it follows that

$$\sum_{k=0}^{n}\frac{(x+y)^k}{k!}\leq\left(\sum_{k=0}^{n}\frac{x^k}{k!}\right)\left(\sum_{k=0}^{n}\frac{y^k}{k!}\right)\leq\sum_{k=0}^{2n}\frac{(x+y)^k}{k!}.$$

But

$$\lim_{n\to\infty}\sum_{k=0}^{n}\frac{(x+y)^k}{k!}=e^{x+y} \quad\text{and}\quad \lim_{n\to\infty}\sum_{k=0}^{2n}\frac{(x+y)^k}{k!}=e^{x+y}.$$

Thus, by the Squeeze Rule and the Product Rule, we deduce that

$$e^xe^y=\lim_{n\to\infty}\left(\sum_{k=0}^{n}\frac{x^k}{k!}\right)\left(\sum_{k=0}^{n}\frac{y^k}{k!}\right)=e^{x+y},$$

as required.

If x and y are not both positive, then we can verify the equation

$$e^xe^y=e^{x+y}$$

by rearranging it so that all the powers are positive (using $e^x=(e^{-x})^{-1}$) and applying the special case just proved. ■

For example, if $x>y>0$, then

$$e^xe^{-y}=e^{x-y}$$

is equivalent to

$$e^x=e^{x-y}e^y,$$

which is true, since $x-y>0$ and $y>0$.

Solutions to the exercises

1.1 **(a)** Using the formula for summing a geometric series, with $a = r = -\frac{1}{3}$, we obtain

$$s_n = \left(-\tfrac{1}{3}\right) + \left(-\tfrac{1}{3}\right)^2 + \left(-\tfrac{1}{3}\right)^3 + \cdots + \left(-\tfrac{1}{3}\right)^n$$
$$= \frac{(-\frac{1}{3})(1-(-\frac{1}{3})^n)}{1-(-\frac{1}{3})}$$
$$= -\tfrac{1}{4}(1-(-\tfrac{1}{3})^n).$$

Since $\{(-\frac{1}{3})^n\}$ is a basic null sequence,

$$\lim_{n\to\infty} s_n = -\tfrac{1}{4},$$

so

$$\sum_{n=1}^{\infty} \left(-\tfrac{1}{3}\right)^n \text{ is convergent, with sum } -\tfrac{1}{4}.$$

(b) In this case,

$$s_n = (-1) + (-1)^2 + (-1)^3 + \cdots + (-1)^n$$
$$= -1 + 1 - 1 + \cdots + (-1)^n.$$

Thus

$$s_n = \begin{cases} -1, & n \text{ odd}, \\ 0, & n \text{ even}. \end{cases}$$

Hence $s_{2k-1} \to -1$ and $s_{2k} \to 0$, so $\{s_n\}$ is divergent, by the First Subsequence Rule; see Unit AA2, Subsection 4.4. Thus this series is divergent.

(c) Using the formula for summing the first $n+1$ terms of a geometric series, with $a = 1$ and $r = \frac{1}{2}$, we obtain

$$s_n = 1 + \tfrac{1}{2} + (\tfrac{1}{2})^2 + \cdots + (\tfrac{1}{2})^n$$
$$= \frac{1(1-(\frac{1}{2})^{n+1})}{1-\frac{1}{2}} = 2 - (\tfrac{1}{2})^n.$$

Since $\{(\frac{1}{2})^n\}$ is a basic null sequence,

$$\lim_{n\to\infty} s_n = 2,$$

so

$$\sum_{n=0}^{\infty} \left(\tfrac{1}{2}\right)^n \text{ is convergent, with sum 2.}$$

1.2 We have

$$s_1 = \frac{1}{1\times 2} = \frac{1}{2},$$
$$s_2 = \frac{1}{1\times 2} + \frac{1}{2\times 3} = \frac{1}{2} + \frac{1}{6} = \frac{2}{3},$$
$$s_3 = \frac{1}{1\times 2} + \frac{1}{2\times 3} + \frac{1}{3\times 4} = \frac{2}{3} + \frac{1}{12} = \frac{3}{4},$$
$$s_4 = \frac{1}{1\times 2} + \frac{1}{2\times 3} + \frac{1}{3\times 4} + \frac{1}{4\times 5} = \frac{3}{4} + \frac{1}{20} = \frac{4}{5}.$$

1.3 Since

$$\frac{1}{n(n+2)} = \frac{1}{2}\left(\frac{1}{n} - \frac{1}{n+2}\right), \quad \text{for } n = 1, 2, \ldots,$$

we have

$$\sum_{n=1}^{\infty} \frac{1}{n(n+2)} = \frac{1}{2}\sum_{n=1}^{\infty}\left(\frac{1}{n} - \frac{1}{n+2}\right).$$

Thus

$$s_n = \frac{1}{1\times 3} + \frac{1}{2\times 4} + \frac{1}{3\times 5} + \cdots + \frac{1}{n(n+2)}$$
$$= \frac{1}{2}\left(\left(1-\frac{1}{3}\right) + \left(\frac{1}{2}-\frac{1}{4}\right) + \left(\frac{1}{3}-\frac{1}{5}\right) + \cdots + \left(\frac{1}{n} - \frac{1}{n+2}\right)\right).$$

Most of the terms in alternate brackets cancel, leaving

$$s_n = \frac{1}{2}\left(1 + \frac{1}{2} - \frac{1}{n+1} - \frac{1}{n+2}\right).$$

Since $\left\{\dfrac{1}{n+1}\right\}$ and $\left\{\dfrac{1}{n+2}\right\}$ are null sequences,

$$\lim_{n\to\infty} s_n = \tfrac{1}{2}(1+\tfrac{1}{2}) = \tfrac{3}{4},$$

so

$$\sum_{n=1}^{\infty} \frac{1}{n(n+2)} \text{ is convergent, with sum } \tfrac{3}{4}.$$

1.4 The series $\sum\limits_{n=1}^{\infty} \left(-\frac{3}{4}\right)^n$ is a geometric series, with $a = -\frac{3}{4}$ and $r = -\frac{3}{4}$. Since $|-\frac{3}{4}| = \frac{3}{4} < 1$, the series is convergent, with sum

$$\frac{-\frac{3}{4}}{1-(-\frac{3}{4})} = -\tfrac{3}{7}.$$

The series $\sum\limits_{n=1}^{\infty} \dfrac{1}{n(n+1)}$ is convergent, with sum 1; see Subsection 1.2. Hence, by the Combination Rules,

$$\sum_{n=1}^{\infty}\left(\left(-\tfrac{3}{4}\right)^n - \frac{2}{n(n+1)}\right) \text{ is convergent,}$$

with sum $-\frac{3}{7} - (2\times 1) = -\frac{17}{7}$.

1.5 Let

$$a_n = \frac{(-1)^{n+1}n^2}{2n^2+1}; \quad \text{then} \quad |a_n| = \frac{n^2}{2n^2+1}.$$

By the Combination Rules for sequences,

$$\lim_{n\to\infty} \frac{n^2}{2n^2+1} = \lim_{n\to\infty} \frac{1}{2+1/n^2} = \tfrac{1}{2} \neq 0,$$

so the sequence $\left\{\dfrac{n^2}{2n^2+1}\right\}$ is not null.

Hence, by the Non-null Test,

$$\sum_{n=1}^{\infty} \frac{(-1)^{n+1}n^2}{2n^2+1} \text{ is divergent.}$$

1.6 This is a geometric series with $a = 1$ and $r = -\frac{4}{5}$. Since $|-\frac{4}{5}| = \frac{4}{5} < 1$, the series is convergent, with sum

$$\frac{1}{1-(-\frac{4}{5})} = \frac{5}{9}.$$

1.7 For $n = 1, 2, \ldots$,

$$\frac{1}{2n-1} - \frac{1}{2n+1} = \frac{(2n+1)-(2n-1)}{(2n-1)(2n+1)} = \frac{2}{(2n-1)(2n+1)} = \frac{2}{4n^2-1},$$

as required.

Thus the given series is a telescoping series and the nth partial sum s_n is given by

$$\begin{aligned} s_n &= \sum_{k=1}^{n} \frac{1}{4k^2-1} = \frac{1}{2}\sum_{k=1}^{n}\left(\frac{1}{2k-1} - \frac{1}{2k+1}\right) \\ &= \frac{1}{2}\left(\left(\frac{1}{1}-\frac{1}{3}\right) + \left(\frac{1}{3}-\frac{1}{5}\right) + \cdots \right. \\ &\qquad \left. + \left(\frac{1}{2n-1} - \frac{1}{2n+1}\right)\right) \\ &= \frac{1}{2}\left(1 - \frac{1}{2n+1}\right). \end{aligned}$$

Since $\{1/(2n+1)\}$ is a null sequence,

$$\lim_{n\to\infty} s_n = \frac{1}{2},$$

so the series is convergent, with sum $\frac{1}{2}$.

1.8 (a) The series $\sum_{n=1}^{\infty} \left(\frac{4}{5}\right)^n$ is a geometric series, with $a = r = \frac{4}{5}$. Since $\frac{4}{5} < 1$, the series is convergent, with sum

$$\frac{\frac{4}{5}}{1-\frac{4}{5}} = 4.$$

The series $\sum_{n=1}^{\infty} \frac{1}{n(n+2)}$ is convergent, with sum $\frac{3}{4}$; see the solution to Exercise 1.3.

Hence, by the Combination Rules,

$$\sum_{n=1}^{\infty} \left(\left(\tfrac{4}{5}\right)^n + \frac{4}{n(n+2)}\right) \text{ is convergent,}$$

with sum $4 + \left(4 \times \frac{3}{4}\right) = 7$.

(b) Since $\{(\frac{1}{2})^n\}$ is a basic null sequence, the sequence $\{1 + (\frac{1}{2})^n\}$ is convergent, with limit $1 \neq 0$. Hence, by the Non-null Test,

$$\sum_{n=1}^{\infty} \left(1 + \left(\tfrac{1}{2}\right)^n\right) \text{ is divergent.}$$

2.1 Let

$$s_n = 1 + \frac{1}{2} + \frac{1}{3} + \cdots + \frac{1}{n}$$

and

$$t_n = 1 + \frac{1}{2^2} + \frac{1}{3^2} + \cdots + \frac{1}{n^2}.$$

The values of these partial sums are given below to 2 decimal places.

n	1	2	3	4	5	6	7	8
s_n	1	1.5	1.83	2.08	2.28	2.45	2.59	2.72
t_n	1	1.25	1.36	1.42	1.46	1.49	1.51	1.53

The sequences are plotted below.

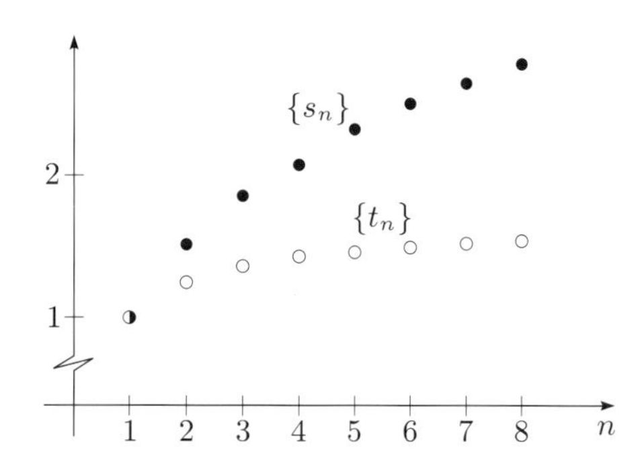

2.2 (a) We guess that this series is dominated by $\sum 1/n^3$. We have

$$n^3 + n \geq n^3, \quad \text{for } n = 1, 2, \ldots,$$

so

$$\frac{1}{n^3+n} \leq \frac{1}{n^3}, \quad \text{for } n = 1, 2, \ldots.$$

Since $\sum_{n=1}^{\infty} 1/n^3$ is convergent, we deduce from the Comparison Test that

$$\sum_{n=1}^{\infty} \frac{1}{n^3+n} \text{ is convergent.}$$

(b) Let

$$a_n = \frac{1}{n+\sqrt{n}}, \quad n = 1, 2, \ldots.$$

We guess that the terms of this series behave like $1/n$ and we know that $\sum 1/n$ is divergent. There is no obvious direct comparison series, so we use the Limit Comparison Test with

$$b_n = \frac{1}{n}, \quad n = 1, 2, \ldots.$$

We obtain

$$\begin{aligned} \frac{a_n}{b_n} &= \left(\frac{1}{n+\sqrt{n}}\right)\left(\frac{n}{1}\right) \\ &= \frac{n}{n+\sqrt{n}} \\ &= \frac{1}{1+1/\sqrt{n}} \to 1 \neq 0. \end{aligned}$$

Since $\sum_{n=1}^{\infty} 1/n$ is divergent, we deduce by the Limit Comparison Test that

$$\sum_{n=1}^{\infty} \frac{1}{n+\sqrt{n}} \text{ is divergent.}$$

(c) Let

$$a_n = \frac{n+4}{2n^3-n+1}, \quad n = 1, 2, \ldots.$$

We guess that the terms of this series behave like $n/n^3 = 1/n^2$. There is no obvious direct comparison series, so we use the Limit Comparison Test with

$$b_n = \frac{1}{n^2}, \quad n = 1, 2, \ldots.$$

We obtain

$$\begin{aligned}\frac{a_n}{b_n} &= \left(\frac{n+4}{2n^3-n+1}\right)\left(\frac{n^2}{1}\right)\\ &= \frac{n^3+4n^2}{2n^3-n+1}\\ &= \frac{1+4/n}{2-1/n^2+1/n^3} \to \tfrac{1}{2} \neq 0.\end{aligned}$$

Since $\sum_{n=1}^{\infty} 1/n^2$ is convergent, we deduce by the Limit Comparison Test that

$$\sum_{n=1}^{\infty} \frac{n+4}{2n^3-n+1} \text{ is convergent.}$$

(d) We guess that this series is dominated by $\sum 1/n^3$. Indeed, since

$$0 \leq \cos^2(2n) \leq 1, \quad \text{for } n = 1, 2, \ldots,$$

we have

$$0 \leq \frac{\cos^2(2n)}{n^3} \leq \frac{1}{n^3}, \quad \text{for } n = 1, 2, \ldots.$$

Since $\sum_{n=1}^{\infty} 1/n^3$ is convergent, we deduce by the Comparison Test that

$$\sum_{n=1}^{\infty} \frac{\cos^2(2n)}{n^3} \text{ is convergent.}$$

2.3 **(a)** Let

$$a_n = \frac{n^3}{n!}, \quad n = 1, 2, \ldots;$$

then

$$\begin{aligned}\frac{a_{n+1}}{a_n} &= \left(\frac{(n+1)^3}{(n+1)!}\right)\left(\frac{n!}{n^3}\right)\\ &= \frac{(n+1)^3}{(n+1)n^3}\\ &= \frac{n^2+2n+1}{n^3} = \frac{1}{n} + \frac{2}{n^2} + \frac{1}{n^3}.\end{aligned}$$

Hence, by the Combination Rules for sequences,

$$\frac{a_{n+1}}{a_n} \to 0 \text{ as } n \to \infty.$$

Thus, by the Ratio Test,

$$\sum_{n=1}^{\infty} \frac{n^3}{n!} \text{ is convergent.}$$

(b) Let

$$a_n = \frac{n^2 2^n}{n!}, \quad n = 1, 2, \ldots;$$

then

$$\begin{aligned}\frac{a_{n+1}}{a_n} &= \left(\frac{(n+1)^2 2^{n+1}}{(n+1)!}\right)\left(\frac{n!}{n^2 2^n}\right)\\ &= \frac{2(n+1)^2}{(n+1)n^2} = 2\left(\frac{1}{n} + \frac{1}{n^2}\right).\end{aligned}$$

Hence, by the Combination Rules for sequences,

$$\frac{a_{n+1}}{a_n} \to 0 \text{ as } n \to \infty.$$

Thus, by the Ratio Test,

$$\sum_{n=1}^{\infty} \frac{n^2 2^n}{n!} \text{ is convergent.}$$

(c) Let

$$a_n = \frac{(2n)!}{n^n}, \quad n = 1, 2, \ldots;$$

then

$$\begin{aligned}\frac{a_{n+1}}{a_n} &= \left(\frac{(2(n+1))!}{(n+1)^{n+1}}\right)\left(\frac{n^n}{(2n)!}\right)\\ &= \frac{(2n+2)!\, n^n}{(n+1)^{n+1}(2n)!}\\ &= \frac{(2n+2)(2n+1)n^n}{(n+1)^{n+1}}\\ &= \frac{2(2n+1)n^n}{(n+1)^n} = \frac{4n+2}{(1+1/n)^n}.\end{aligned}$$

Now

$$\lim_{n\to\infty} (1+1/n)^n = e \quad \text{and} \quad \lim_{n\to\infty} \frac{1}{4n+2} = 0,$$

so $\{(1+1/n)^n/(4n+2)\}$ is null, by the Product Rule.

We deduce by the Reciprocal Rule that

$$\frac{a_{n+1}}{a_n} \to \infty \quad \text{as } n \to \infty;$$

see Unit AA2, Section 4.

Hence, by the Ratio Test,

$$\sum_{n=1}^{\infty} \frac{(2n)!}{n^n} \text{ is divergent.}$$

(Alternatively, note that

$$\begin{aligned}\frac{(2n)!}{n^n} &\geq \left(\frac{2n}{n}\right)\left(\frac{2n-1}{n}\right)\cdots\left(\frac{n+1}{n}\right)\\ &\geq 1,\end{aligned}$$

so, by the Non-null Test,

$$\sum_{n=1}^{\infty} \frac{(2n)!}{n^n} \text{ is divergent.)}$$

2.4 (a) We guess that this series is dominated by $\sum 1/n^2$. Indeed, since

$$0 \le 1/n \le \frac{\pi}{2}, \quad \text{for } n = 1, 2, \ldots,$$

we have

$$0 \le \cos(1/n) \le 1, \quad \text{for } n = 1, 2, \ldots.$$

Hence, for $n = 1, 2, \ldots,$

$$0 \le \frac{\cos(1/n)}{2n^2+3} \le \frac{1}{2n^2+3} \le \frac{1}{n^2}.$$

Since $\sum_{n=1}^{\infty} 1/n^2$ is a basic convergent series, we deduce by the Comparison Test that

$$\sum_{n=1}^{\infty} \frac{\cos(1/n)}{2n^2+3} \text{ is convergent.}$$

(b) Let

$$a_n = \frac{n^2}{2n^3 - n}, \quad n = 1, 2, \ldots.$$

We guess that the terms of this series behave like $n^2/n^3 = 1/n$. There is no obvious direct comparison series, so we use the Limit Comparison Test with

$$b_n = \frac{1}{n}, \quad n = 1, 2, \ldots.$$

By the Combination Rules for sequences,

$$\frac{a_n}{b_n} = \left(\frac{n^2}{2n^3 - n}\right)\left(\frac{n}{1}\right) = \frac{1}{2 - 1/n^2} \to \tfrac{1}{2} \ne 0.$$

Since $\sum_{n=1}^{\infty} 1/n$ is a basic divergent series, we deduce by the Limit Comparison Test that

$$\sum_{n=1}^{\infty} \frac{n^2}{2n^3 - n} \text{ is divergent.}$$

(c) Let

$$a_n = \frac{\sqrt{2n}}{4n^3 + n + 2}, \quad n = 1, 2, \ldots.$$

We guess that the terms of this series behave like $\sqrt{n}/n^3 = 1/n^{5/2}$. There is no obvious direct comparison series, so we use the Limit Comparison Test with

$$b_n = \frac{1}{n^{5/2}}, \quad n = 1, 2, \ldots.$$

By the Combination Rules for sequences,

$$\frac{a_n}{b_n} = \left(\frac{\sqrt{2n}}{4n^3 + n + 2}\right)\left(\frac{n^{5/2}}{1}\right) = \frac{\sqrt{2}\, n^3}{4n^3 + n + 2} = \frac{\sqrt{2}}{4 + 1/n^2 + 2/n^3} \to \tfrac{1}{4}\sqrt{2} \ne 0.$$

Since $\sum_{n=1}^{\infty} 1/n^{5/2}$ is a basic convergent series, we deduce by the Limit Comparison Test that

$$\sum_{n=1}^{\infty} \frac{\sqrt{2n}}{4n^3 + n + 2} \text{ is convergent.}$$

(d) Let

$$a_n = \frac{(n+1)^5}{2^n}, \quad n = 1, 2, \ldots.$$

We guess that the terms of this series behave like $n^5/2^n$, so we use the Limit Comparison Test with

$$b_n = \frac{n^5}{2^n}, \quad n = 1, 2, \ldots.$$

By the Combination Rules for sequences,

$$\frac{a_n}{b_n} = \left(\frac{(n+1)^5}{2^n}\right)\left(\frac{2^n}{n^5}\right) = \left(\frac{n+1}{n}\right)^5 = \left(1 + \frac{1}{n}\right)^5 \to 1 \ne 0.$$

Since $\sum_{n=1}^{\infty} n^5/2^n$ is a basic convergent series, we deduce by the Limit Comparison Test that

$$\sum_{n=1}^{\infty} \frac{(n+1)^5}{2^n} \text{ is convergent.}$$

(Alternatively, we can either note that

$$\sum_{n=1}^{\infty} \frac{(n+1)^5}{2^n} = \sum_{n=2}^{\infty} \frac{n^5}{2^{n-1}} = 2\sum_{n=2}^{\infty} \frac{n^5}{2^n},$$

and use the Multiple Rule, or use the Ratio Test:

$$\frac{a_{n+1}}{a_n} = \left(\frac{(n+2)^5}{2^{n+1}}\right)\left(\frac{2^n}{(n+1)^5}\right) = \frac{1}{2}\left(\frac{1 + 2/n}{1 + 1/n}\right)^5,$$

which converges to $\frac{1}{2} < 1$.)

(e) There is no obvious basic series to compare with, so we try the Ratio Test. Let

$$a_n = \frac{n^2 3^n}{n!}, \quad n = 1, 2, \ldots;$$

then

$$\frac{a_{n+1}}{a_n} = \left(\frac{(n+1)^2 3^{n+1}}{(n+1)!}\right)\left(\frac{n!}{n^2 3^n}\right) = \frac{3(n+1)}{n^2} = 3\left(\frac{1}{n} + \frac{1}{n^2}\right).$$

By the Combination Rules for sequences,

$$\frac{a_{n+1}}{a_n} \to 0 < 1,$$

so, by the Ratio Test,

$$\sum_{n=1}^{\infty} \frac{n^2 3^n}{n!} \text{ is convergent.}$$

(f) Again, we try the Ratio Test. Let

$$a_n = \frac{(n!)^2}{(2n)!}, \quad n = 1, 2, \dots;$$

then

$$\begin{aligned}\frac{a_{n+1}}{a_n} &= \left(\frac{((n+1)!)^2}{(2n+2)!}\right)\left(\frac{(2n)!}{(n!)^2}\right)\\ &= \frac{(n+1)^2}{(2n+2)(2n+1)}\\ &= \frac{n+1}{2(2n+1)} = \frac{1+1/n}{2(2+1/n)}.\end{aligned}$$

By the Combination Rules for sequences,

$$\frac{a_{n+1}}{a_n} \to \tfrac{1}{4} < 1.$$

Hence, by the Ratio Test,

$$\sum_{n=1}^{\infty} \frac{(n!)^2}{(2n)!} \text{ is convergent.}$$

2.5 (a) Let

$$a_n = \frac{2^n n!}{n^n}, \quad n = 1, 2, \dots;$$

then

$$\begin{aligned}\frac{a_{n+1}}{a_n} &= \left(\frac{2^{n+1}(n+1)!}{(n+1)^{n+1}}\right)\left(\frac{n^n}{2^n n!}\right)\\ &= \frac{2(n+1)\, n^n}{(n+1)^{n+1}}\\ &= \frac{2n^n}{(n+1)^n} = \frac{2}{(1+1/n)^n}.\end{aligned}$$

Thus

$$\frac{a_{n+1}}{a_n} \to \frac{2}{e} < 1,$$

so, by the Ratio Test,

$$\sum_{n=1}^{\infty} \frac{2^n n!}{n^n} \text{ is convergent.}$$

(b) If $a_n = 3^n n!/n^n$, then a similar calculation yields

$$\frac{a_{n+1}}{a_n} \to \frac{3}{e} > 1,$$

so, by the Ratio Test,

$$\sum_{n=1}^{\infty} \frac{3^n n!}{n^n} \text{ is divergent.}$$

Remark A similar argument shows that if $a_n = x^n n!/n^n$, where $x > 0$, then the series

$$\sum_{n=1}^{\infty} \frac{x^n n!}{n^n}$$

is convergent if $0 < x < e$ and divergent if $x > e$.

If $x = e$, then the series is also divergent, but this is harder to prove. One method is to use the inequality

$$n! > \left(\frac{n+1}{e}\right)^n, \quad \text{for } n = 1, 2, \dots;$$

see Unit AA2, Exercise 5.5.

From this inequality, we deduce that

$$\frac{e^n n!}{n^n} > \left(\frac{n+1}{n}\right)^n = (1+1/n)^n, \quad \text{for } n = 1, 2, \dots.$$

Since $\lim\limits_{n\to\infty} (1+1/n)^n = e \neq 0$, we deduce that $\{e^n n!/n^n\}$ is not a null sequence.

Hence, by the Non-null Test,

$$\sum_{n=1}^{\infty} \frac{e^n n!}{n^n} \text{ is divergent.}$$

3.1 (a) Let

$$a_n = \frac{(-1)^{n+1} n}{n^3+1}, \quad n = 1, 2, \dots;$$

then

$$|a_n| = \frac{n}{n^3+1}, \quad \text{for } n = 1, 2, \dots.$$

Now

$$\frac{n}{n^3+1} \leq \frac{n}{n^3} = \frac{1}{n^2}, \quad \text{for } n = 1, 2, \dots,$$

and $\sum\limits_{n=1}^{\infty} 1/n^2$ is a basic convergent series. Hence, by the Comparison Test,

$$\sum_{n=1}^{\infty} \frac{n}{n^3+1} \text{ is convergent.}$$

By the Absolute Convergence Test, it follows that

$$\sum_{n=1}^{\infty} \frac{(-1)^{n+1} n}{n^3+1} \text{ is convergent.}$$

(b) The series $\sum\limits_{n=0}^{\infty} 1/2^n$ is a convergent geometric series. Thus

$$1 + \frac{1}{2} - \frac{1}{4} + \frac{1}{8} + \frac{1}{16} - \frac{1}{32} + \cdots$$

is absolutely convergent. Hence, by the Absolute Convergence Test,

$$1 + \frac{1}{2} - \frac{1}{4} + \frac{1}{8} + \frac{1}{16} - \frac{1}{32} + \cdots \text{ is convergent.}$$

3.2 The series

$$\frac{1}{2} - \frac{1}{4} - \frac{1}{8} + \frac{1}{16} - \frac{1}{32} - \frac{1}{64} + \cdots$$

is absolutely convergent. Thus it is convergent, by the Absolute Convergence Test. By the infinite form of the Triangle Inequality,

$$\left|\frac{1}{2} - \frac{1}{4} - \frac{1}{8} + \cdots\right| \leq \frac{1}{2} + \frac{1}{4} + \frac{1}{8} + \cdots = 1.$$

Hence the sum lies in the interval $[-1, 1]$.

3.3 **(a)** The sequence $\{(-1)^{n+1}/n^{1/3}\}$ is of the form $\{(-1)^{n+1}b_n\}$, where

$$b_n = 1/n^{1/3}, \quad n = 1, 2, \ldots.$$

Now

1. $1/n^{1/3} \geq 0$, for $n = 1, 2, \ldots$;
2. $\{1/n^{1/3}\}$ is a basic null sequence;
3. $\{1/n^{1/3}\}$ is decreasing (because $\{n^{1/3}\}$ is increasing).

Hence, by the Alternating Test,

$$\sum_{n=1}^{\infty} \frac{(-1)^{n+1}}{n^{1/3}} \text{ is convergent.}$$

(b) The sequence $\{(-1)^{n+1}/(n + n^{1/2})\}$ is of the form $\{(-1)^{n+1}b_n\}$, where

$$b_n = 1/(n + n^{1/2}), \quad n = 1, 2, \ldots.$$

Now

1. $1/(n + n^{1/2}) \geq 0$, for $n = 1, 2, \ldots$;
2. $\{1/(n + n^{1/2})\}$ is a null sequence, by the Squeeze Rule, since
$$0 \leq \frac{1}{n + n^{1/2}} \leq \frac{1}{n}, \quad \text{for } n = 1, 2, \ldots,$$
and $\{1/n\}$ is a basic null sequence;
3. $\{1/(n + n^{1/2}\}$ is decreasing (because $\{n + n^{1/2}\}$ is increasing).

Hence, by the Alternating Test,

$$\sum_{n=1}^{\infty} \frac{(-1)^{n+1}}{n + n^{1/2}} \text{ is convergent.}$$

(c) The sequence $\{(-1)^{n+1}n/(n+2)\}$ is not a null sequence because

$$\left|\frac{(-1)^{n+1}n}{n+2}\right| = \frac{n}{n+2} \to 1 \neq 0.$$

Hence, by the Non-null Test,

$$\sum_{n=1}^{\infty} \frac{(-1)^{n+1}n}{n+2} \text{ is divergent.}$$

3.4 **(a)** We have

$$\frac{5n + 2^n}{3^n} = 5n\left(\tfrac{1}{3}\right)^n + \left(\tfrac{2}{3}\right)^n, \quad \text{for } n = 1, 2, \ldots.$$

Now

$$\sum_{n=1}^{\infty} n\left(\tfrac{1}{3}\right)^n \quad \text{and} \quad \sum_{n=1}^{\infty} \left(\tfrac{2}{3}\right)^n$$

are both basic convergent series. Thus, by the Combination Rules for series,

$$\sum_{n=1}^{\infty} \frac{5n + 2^n}{3^n} \text{ is convergent.}$$

(b) We guess that the terms of the series behave like $1/n^3$, so we use the Limit Comparison Test with

$$b_n = 1/n^3, \quad n = 1, 2, \ldots.$$

We have

$$\frac{a_n}{b_n} = \left(\frac{3}{2n^3 - 1}\right)\left(\frac{n^3}{1}\right) = \frac{3n^3}{2n^3 - 1} \to \tfrac{3}{2} \neq 0.$$

Since $\sum_{n=1}^{\infty} 1/n^3$ is a basic convergent series, we deduce from the Limit Comparison Test that

$$\sum_{n=1}^{\infty} \frac{3}{2n^3 - 1} \text{ is convergent.}$$

(c) We use the Alternating Test.

The sequence $\{(-1)^{n+1}/(n \log_e(n+1))\}$ is of the form $\{(-1)^{n+1}b_n\}$, where

$$b_n = \frac{1}{n \log_e(n+1)}, \quad n = 1, 2, \ldots.$$

Now

1. $1/(n \log_e(n+1)) \geq 0$, for $n = 1, 2, \ldots$;
2. $\{1/(n \log_e(n+1))\}$ is a null sequence, by the Squeeze Rule, since
$$0 \leq \frac{1}{n \log_e(n+1)} \leq \frac{1}{n \log_e 2}, \quad \text{for } n = 1, 2, \ldots,$$
and $\{1/n\}$ is a basic null sequence;
3. $\{1/(n \log_e(n+1))\}$ is decreasing (because $\{n \log_e(n+1)\}$ is increasing).

Hence, by the Alternating Test,

$$\sum_{n=1}^{\infty} \frac{(-1)^{n+1}}{n \log_e(n+1)} \text{ is convergent.}$$

(d) The sequence

$$a_n = \frac{(-1)^{n+1}n^2}{n^2 + 1}, \quad n = 1, 2, \ldots,$$

is not null, since

$$|a_n| = \frac{n^2}{n^2 + 1} = \frac{1}{1 + 1/n^2} \to 1 \neq 0.$$

Hence, by the Non-null Test,

$$\sum_{n=1}^{\infty} a_n \text{ is divergent.}$$

(e) Let

$$a_n = \frac{(-1)^{n+1}n}{n^3 + 5}, \quad n = 1, 2, \ldots;$$

then

$$|a_n| = \frac{n}{n^3 + 5}, \quad \text{for } n = 1, 2, \ldots.$$

Thus

$$|a_n| \leq \frac{n}{n^3} = \frac{1}{n^2}, \quad \text{for } n = 1, 2, \ldots.$$

Since $\sum 1/n^2$ is a basic convergent series, we deduce by the Comparison Test that $\sum_{n=1}^{\infty} |a_n|$ is convergent, so $\sum_{n=1}^{\infty} a_n$ is absolutely convergent.

Hence, by the Absolute Convergence Test,

$$\sum_{n=1}^{\infty} \frac{(-1)^{n+1} n}{n^3+5} \text{ is convergent.}$$

(f) Since $\{n^6/2^n\} = \{n^6(\frac{1}{2})^n\}$ is a basic null sequence, we deduce from the Reciprocal Rule that

$$\frac{2^n}{n^6} \to \infty \quad \text{as } n \to \infty;$$

see Unit AA2, Section 4.

Thus $\{2^n/n^6\}$ is not a null sequence. Hence, by the Non-null Test,

$$\sum_{n=1}^{\infty} \frac{2^n}{n^6} \text{ is divergent.}$$

3.5 (a) We use the Alternating Test.

Let

$$a_n = \frac{(-1)^{n+1}}{1+\sqrt{n}}, \quad n = 1, 2, \ldots.$$

Then $a_n = (-1)^{n+1} b_n$, for $n = 1, 2, \ldots$, where

$$b_n = \frac{1}{1+\sqrt{n}}, \quad n = 1, 2, \ldots,$$

and

1. $1/(1+\sqrt{n}) \geq 0$, for $n = 1, 2, \ldots$;
2. $\{1/(1+\sqrt{n})\}$ is a null sequence;
3. $\{1/(1+\sqrt{n})\}$ is decreasing (because $\{1+\sqrt{n}\}$ is increasing).

Hence, by the Alternating Test,

$$\sum_{n=1}^{\infty} \frac{(-1)^{n+1}}{1+\sqrt{n}} \text{ is convergent.}$$

(b) We guess that this series is dominated by the basic convergent series $\sum 1/n^2$. Indeed,

$$|\sin n| \leq 1, \quad \text{for } n = 1, 2, \ldots,$$

so

$$\left|\frac{\sin n}{n^2}\right| \leq \frac{1}{n^2}, \quad \text{for } n = 1, 2, \ldots.$$

Thus $\sum_{n=1}^{\infty} \frac{\sin n}{n^2}$ is absolutely convergent, by the Comparison Test, and hence convergent by the Absolute Convergence Test.

(c) Let

$$a_n = \frac{(-1)^{n+1} n!}{n^4+3}, \quad n = 1, 2, \ldots.$$

Now, by the Reciprocal Rule and the Sum Rule for sequences,

$$|a_n| = \frac{n!}{n^4+3} \to \infty \quad \text{as } n \to \infty.$$

Thus, by the Non-null Test,

$$\sum_{n=1}^{\infty} \frac{(-1)^{n+1} n!}{n^4+3} \text{ is divergent.}$$

(d) Let

$$a_n = \frac{n+2^n}{3^n+5}, \quad n = 1, 2, \ldots.$$

We guess that a_n behaves like $2^n/3^n = (\frac{2}{3})^n$, so we use the Limit Comparison Test with

$$b_n = \left(\tfrac{2}{3}\right)^n, \quad n = 1, 2, \ldots.$$

We have

$$\begin{aligned}\frac{a_n}{b_n} &= \left(\frac{n+2^n}{3^n+5}\right)\left(\frac{3}{2}\right)^n \\ &= \frac{3^n(n+2^n)}{2^n(3^n+5)} \\ &= \frac{n/2^n+1}{1+5/3^n} \to 1 \neq 0,\end{aligned}$$

since $\{n(\frac{1}{2})^n\}$ and $\{(\frac{1}{3})^n\}$ are basic null sequences.

Now $\sum_{n=1}^{\infty} \left(\frac{2}{3}\right)^n$ is a basic convergent series, so

$$\sum_{n=1}^{\infty} \frac{n+2^n}{3^n+5} \text{ is convergent,}$$

by the Limit Comparison Test.

(e) We use the Alternating Test.

Let

$$a_n = \frac{(-1)^{n+1} n}{n^2+2}, \quad n = 1, 2, \ldots.$$

Then a_n is of the form $(-1)^{n+1} b_n$, where

$$b_n = \frac{n}{n^2+2}, \quad n = 1, 2, \ldots,$$

and

1. $n/(n^2+2) \geq 0$, for $n = 1, 2, \ldots$;
2. $\{n/(n^2+2)\}$ is a null sequence;
3. $\{n/(n^2+2)\}$ is decreasing.

To show that the sequence $\{b_n\} = \{n/(n^2+2)\}$ is decreasing, we show that the sequence

$$\frac{1}{b_n} = \frac{n^2+2}{n}, \quad n = 1, 2, \ldots,$$

is increasing by using Unit AA2, Strategy 1.1. We have

$$\frac{n^2+2}{n} = n + \frac{2}{n},$$

so

$$\begin{aligned}\frac{1}{b_{n+1}} - \frac{1}{b_n} &= \left(n+1+\frac{2}{n+1}\right) - \left(n+\frac{2}{n}\right) \\ &= 1 - \frac{2}{n(n+1)} \geq 0.\end{aligned}$$

Thus $\{1/b_n\}$ is increasing, as required.

Hence, by the Alternating Test,

$$\sum_{n=1}^{\infty} \frac{(-1)^{n+1} n}{n^2+2} \text{ is convergent.}$$

3.6 Since $3n-2<3n$ and $3n-1<3n$, we have

$$\frac{1}{3n-2}+\frac{1}{3n-1}-\frac{1}{3n}>\frac{1}{3n}+\frac{1}{3n}-\frac{1}{3n}$$
$$=\frac{1}{3n},\quad \text{for } n=1,2,\ldots.$$

Thus

$$\begin{aligned}s_{3n}&=\left(1+\frac{1}{2}-\frac{1}{3}\right)+\left(\frac{1}{4}+\frac{1}{5}-\frac{1}{6}\right)+\cdots\\&\quad+\left(\frac{1}{3n-2}+\frac{1}{3n-1}-\frac{1}{3n}\right)\\&>\frac{1}{3}+\frac{1}{6}+\cdots+\frac{1}{3n}\\&=\frac{1}{3}\left(1+\frac{1}{2}+\cdots+\frac{1}{n}\right).\end{aligned}$$

We know that the sequence

$$\left\{1+\frac{1}{2}+\cdots+\frac{1}{n}\right\}\text{ is unbounded;}$$

see Subsection 2.1, Frame 2.

Thus $\{s_{3n}\}$ is also unbounded, so $\{s_n\}$ is divergent.
Hence the series

$$1+\frac{1}{2}-\frac{1}{3}+\frac{1}{4}+\frac{1}{5}-\frac{1}{6}+\cdots\text{ is divergent.}$$

Index